Cold and Chilled
Storage Technology

Edited by

CLIVE V.J. DELLINO
Managing Director
Tempco Engineering Services
Bedford

Blackie
Glasgow and London

Van Nostrand Reinhold
New York

Blackie and Son Ltd
Bishopbriggs, Glasgow G64 2NZ
and
7 Leicester Place, London WC2H 7BP

Published in the United States of America by
Van Nostrand Reinhold
115 Fifth Avenue
New York, New York 10003

Distributed in Canada by
Nelson Canada
1120 Birchmount Road
Scarborough, Ontario M1K 5G4, Canada

16 15 14 13 12 11 10 9 8 7 6 5 4 3 2 1

British Library Cataloguing in Publication Data

Cold and chilled storage technology.
  1. Food. Cold storage
  I. Dellino, Clive V.J.
  664'.02852

  ISBN 0-216-92502-9

Library of Congress Cataloging-in-Publication Data

Cold and chilled storage technology.

  Bibliography: p.
  Includes index.
  1. Cold storage. I. Dellino, Clive V.J.
TP372.2.C64 1988 664'.02852   88-5488
ISBN 0-442-20673-9

Phototypesetting by Thomson Press (India) Ltd., New Delhi
Printed in Great Britain by Thomson Litho Ltd, East Kilbride, Scotland

Preface

When I was asked to act as Editor of this book, it was suggested that after thirty-three years in the industry I must have *some* friends who know what they are talking about. I hope the reader will agree that the contributors to the various chapters not only know their subjects, but also have presented summaries of their specialist knowledge in a way that interlocks into the jigsaw picture of our industry. There may well be gaps, but this is due solely to the need to keep the volume to a manageable size. The choice of what to include and what to omit was mine, and I ask the reader's understanding if some aspect is missing or inadequately covered.

The first four chapters are concerned primarily with the basic functions and operation of the various types of temperature controlled facilities in the food distribution chain. The remaining chapters consider specific technical areas and equipment that make up the unique nature of temperature controlled warehousing and goods handling. Chapter 3 however, on Controlled Atmosphere Storage, itself includes brief information on the appropriate equipment specific to this application, whilst that on Fully Automated cold stores comes in the later 'equipment' chapters as it is not specific to a storage function or temperature regime.

In the selection of authors, I have drawn on those with experience in the UK, USA and Continental Europe. Our industry is becoming more integrated in international terms and as development takes place, we are all facing similar problems and finding similar solutions. For example, this book was written during a period of great activity in the UK contracting industry when a pronounced development of the composite, multi-temperature distribution depots for the major retail chains was taking place. The impact of the new requirements of these sites is reflected in the content of the relevant chapters and is equally applicable outside the UK.

I hope that this book will be a useful summary of the current situation in our industry and that it will be a valuable reference source for its readers. Finally, I wish to record my sincere appreciation of the efforts made by old and new friends who have so generously given of their time and experience.

CVJD

Contributors

B. Barnes, Bryan Barnes Associates, Consultants, Cheltenham. Previously with Bird's Eye Wall's.

D.J. Bishop, Managing Director, David Bishop Instruments Ltd, Heathfield, East Sussex.

F.W. Carr, Managing Director, Harry Carr Ltd, Grimsby.

N.P. Daniels, Project Engineer, Energy Management and Technology Division, W S Atkins Ltd, Epsom.

A.F. Harvey, Consultant Adviser (Marketing), Lansing Ltd, Basingstoke.

H.M. Hunter, Sales Director, Refrigeration Division, APV Baker Ltd, Dartford.

A.O. Page, Project Engineer, Energy Management and Technology Division, W S Atkins Ltd, Epsom.

B.A. Russell, Managing Director, United Cold Store Construction Group, London.

R. Showell, Sales Director, Bar Productions (Bromsgrove) Ltd, Bromsgrove.

R. Taylor, Regional Sales Manager, Harnischfeger Engineers Inc., Milwaukee, Wisconsin. Previously Integrated Systems Manager, Materials Handling Division, Interlake Companies Inc., Illinois.

C.W. Toole, General Manager, Technical Centre UK, Frigoscandia Ltd, Hoddesdon, Hertfordshire.

W. van der Spek, Managing Director, Envirodoor Markus Ltd, Hull.

Contents

4. Energy conservation   99
N.P. DANIELS and A.O. PAGE

5. Store insulation   134

B.A. RUSSELL

6. Refrigeration plant   177
H.M. HUNTER

7. Electrical Installations   200
F.W. CARR

Abbreviations

AGV	automated guided vehicle
APR	adjustable pallet racking
ASHRAE	American Society of Heating, Refrigeration and Air Conditioning Engineers
AS/RS	automatic refrigeration and storage system
BS	British Standard
CA	controlled atmosphere
CFC	chlorofluorocarbon
COP	coefficient of performance
CPU	central processor unit
CRT	cathode ray tube (monitor)
EEC	European Community
EHDFRA	extra heavy density fire retardant additive
EPROM	erasable programmable read-only memory
EPS	expanded polystyrene
FDA	U.S. Food and Drug Authority
FIFO	first in, first out
FILO	first in, last out
FLT	fork lift truck
GRP	glass reinforced plastic
HDFRA	high density fire retardant additive
IARW	International Association of Refrigerated Warehouses
I/O	input/ouput
ISO	International Standards Organisation
LED	light-emitting diode
LEL	lower explosive limit
LIFO	last in, first out
NAPR	narrow aisle adjustable pallet racking
NFFA	National Frozen Food Association
P&D	place and dispatch
PIR	polyisocyanurate
PMAPR	powered mobile adjustable pallet racking
PMNAPR	powered mobile narrow aisle pallet racking
PU	polyurethane foam
PUR	polyurethane rigid foam
PVC	polyvinylchloride
R11, R12, R22, R40 R502, R717	refrigerants
REM	rack entry module
SDFERA	standard density fire retardant additive
SEMA	Storage Equipment Manufacturing Association
SKU	stock-keeping unit
ULO	ultra-low oxygen
VNA	very narrow aisle
WCC	works control computer

Units of measurement

In most chapters, metric units of measurement are used. In the final chapter, following the resident country of the author, imperial units are used and approximate metric equivalents are provided in parentheses.

For the convenience of the reader, some conversion factors between metric and imperial units and some equivalent values are given here.

Temperature

$x°F$ (Fahrenheit) $= \frac{5}{9}(x - 32)°C$

$y°C$ (Centigrade, Celsius) $= (\frac{9}{5}y + 32)°F$

Weight

1 pound (lb) $= 0.45\,kg$

*y ton $= 1.02y$ tonne

1 kilogramme (kg) $= 2.2\,lb$

1 tonne $= 1000\,kg$

1 kg $= 1000\,g$

Centigrade	Fahrenheit	Kilogramme	Pound
− 40	− 40	0.1	0.22
− 35	− 31	0.2	0.44
− 30	− 22	0.3	0.66
− 25	− 13	0.4	0.88
− 20	− 4	0.5	1.10
− 15	+ 5	0.6	1.32
− 10	+ 14	0.7	1.54
− 5	+ 23	0.8	1.76
0	+ 32	0.9	1.98
+ 5	+ 41	1.0	2.20
+ 10	+ 50	10	22.0
+ 15	+ 59	50	110
+ 20	+ 68	100	220
+ 25	+ 77	150	330
+ 30	+ 86	200	440
+ 35	+ 95	250	550
+ 40	+ 104	300	660
+ 45	+ 113	350	770
+ 50	+ 122	400	880
+ 60	+ 140	450	990
+ 70	+ 158	500	1100
+ 80	+ 176	600	1320
+ 90	+ 194	700	1540
+ 100	+ 212	800	1760
+ 150	+ 302	900	1980
+ 200	+ 392	1000	2200
+ 250	+ 482		
+ 300	+ 572		

*The conversion factor between tons and tonnes is almost unity and values in tons/tonnes are used interchangeably.

Distance

1 inch (in) = 2.54 centimetres (cm) = 25.4 millimetres (mm)
1 foot (ft) = 0.3048 m
1 cm = 0.3937 inch
1 metre (m) = 3.28 feet
1 m = 100 cm = 1000 mm

mm	inches	metres	feet
0.5	0.02	0.5	1.64
1	0.04	1	3.28
1.5	0.06	1.5	4.92
2	0.08	2	6.56
2.5	0.10	2.5	8.20
3	0.12	3	9.84
3.5	0.14	3.5	11.48
4	0.16	4	13.12
4.5	0.18	4.5	14.76
5	0.20	5	16.40
10	0.39	6	19.68
20	0.79	7	22.96
30	1.18	8	26.24
40	1.56	9	29.52
50	1.95	10	32.80
100	3.94	15	49.2
150	5.90	20	65.6
200	7.87	25	82.0
250	9.75	30	98.4
300	11.81	35	114.8
350	13.78	40	131.2
400	15.75	45	147.6
450	17.72	50	164.0
500	19.68	55	180.4
550	21.65	60	196.8
600	23.62	65	213.2
		70	229.6
		75	246.0
		80	262.4
		85	278.8
		90	295.2
		95	311.6
		100	328.0

1 Bulk stores and associated services

C.W. TOOLE

1.1 Introduction

The technique of preserving food by the use of low temperature dates from the eighteenth century with the first use of ice houses at least as early as 1750. These structures, which were common among wealthier members of society at that time, consisted of brick-lined pits or wells below ground level in which ice, harvested from local lakes, was stored. The later 1800s saw the construction of purpose-built commercial cold stores which were generally to be found at the major seaports of Europe and America for the purpose of storing carcass meat, dairy and fruit products.

Most of these structures were built on small, valuable areas of land, and multistorey buildings were accepted as the best design available. As well as the expense of land, there was, of course, another good reason why the internal height of these chambers was about 2.5 m, namely, that pallets and lift trucks had not yet been developed for commercial applications. The refrigeration systems were, in the main, a development of those to be found in the marine industry, being mostly brine-circulated secondary systems of one kind or another.

World War II saw the first dramatic change in refrigerated warehousing. The use of the pallet was developed to move ammunition and other supplies, and proved to be a very valuable tool for all forms of product movement and storage at that time. Lift trucks were also developed to improve the handling of heavy loads and subsequently to achieve better utilization of storage space.

In the twenty or so years following World War II, and mainly as a direct result of the development of the frozen food industry, the bulk cold store simply provided a means of holding large tonnages of seasonal products, such as vegetables, ice-cream, butter, meat, and fish. These products were ultimately to be joined by poultry, confectionery, and other added-value products.

In those years, the demands on the bulk store were confined mostly to receiving the product from the processor at the loading bay, checking the units or items delivered and then, with the use of assembled convertors, stacking the product inside the cold store four or five pallet units high. In the main, the product came into the cold store in large tonnages (or lots, to use the modern term), and usually also went out in fairly large consignments back to the processor for further processing or packaging.

Few, if any, additional services were provided for the customer at the cold store; those which existed were generally limited to facilities for freezing either in air blast tunnels or compact plate freezers. The majority of customers did not require additional services to be carried out at the cold store; the main requirement was for correct in-store temperatures, good reliable handling practices, and accurate documentation.

The main change to take place in the bulk store in recent years has been the introduction of multitemperature chambers with humidity control for specific products. Enclosed loading bays afford better environmental protection from the prevailing atmospheric conditions, and generally provide better discharge and loading facilities.

The high cost of transportation, combined with other cost factors, has resulted in a move to even greater reliance upon the additional services provided by the cold store. In many instances, the simple freezing service has given way to full process facility, where the customer delivers the raw material to the site and the cold store carries out a full process on the product. The customer will, in most cases, require his representative to be on-site in order to ensure that his quality standards are met at all times.

However, the escalating costs of utilities such as electricity, water, and gas, and ever-increasing labour costs, have on the one hand forced the public cold store operator to improve his techniques and operating facilities, and on the other hand have forced the customer to limit the volume of individual products held in the store so that storage costs are at all times kept to a minimum.

1.2 Building construction

Once a suitable site upon which to build a cold store has been chosen, the modern cold store development must meet a number of specific requirements that will allow it to accommodate the ever-changing demands placed on it by its customers.

In general, it needs to be located near to a major road network with suitable access and within easy reach of attendant commercial support, in the form of specialized contractors as well as general, permanent personnel. It is essential that all the main utilities such as electricity, water, and gas are readily available; otherwise the premium for providing them can be an unnecessary cost penalty.

The site itself should obviously be as level as possible to avoid unnecessary movement of soil; the land should also be stable and as structurally sound as possible to avoid unnecessary expenditure in establishing a structural base.

The most important element in an industrial building of this nature consists of the ground works and the main floor slab itself. It is generally accepted that loading bay facilities should be provided, incorporating the means for discharging from road transport vehicles directly into an enclosed environment on the loading bay. In order to achieve this level of sophistication, it is likely that the loading bay height will have to be approximately 1.4 m above the traffic yard level. In order to accommodate the varying heights of modern trunker road vehicles, dock levellers are introduced to make the connection between the loading bay and the vehicle as perfect as possible.

The chamber configuration within a bulk cold store will depend entirely upon the anticipated business available in that particular area. It should, however, be as flexible as possible because the available business is likely to change frequently due to changing customer demands and market trends, and the need to provide in some cases a variety of storage facilities for one particular customer with a range of different products.

Over the years, a reasonable standard module plan has been used which can provide the operator with the best use of the space constructed and the ability to change the internal layout of each chamber as the business demands. The conventional insulated envelope is (in the UK) usually constructed around an external steel frame with internal insulated panels which are in turn protected by weatherproof sheeting over the roof and all or part of the vertical walls.

Earlier designs provided an internal structure, usually using concrete columns and steel trusses supporting the top roof which in turn was insulated; the external insulated panels provided the weatherproof cladding for the structure. In many parts of the world the external system is still used. Over the years, technological improvements in all aspects of cold store construction have greatly added to the constructors' ability to maintain a reasonable cost level for building new bulk cold stores on green-field sites. The internal height of the cold store is important, and serious consideration must be given at the planning stage to the ultimate working height within each chamber. It is important today that consideration is given to the variety of alternative stacking techniques that are available to the industry, and the use of internal racking in various forms is becoming more and more important at the planning stage.

The following sections describe the basic concept of in-store layout using the various alternative racking systems available to the industry.

1.3 Traffic flow

When planning a bulk cold store, it is normal to consider the following elements in assessing the basic parameters on which the design will be formulated:

- Product types
- Tonnages

- Lot sizes
- Expected turnover
- Seasonal fluctuations
- Packaging
- Arrival temperature
- Transport used e.g. road, rail, or both—in and out
- Pallets used in conjunction with the transport
- Pallet size and design
- Ancillary services, such as check weighing, marking, and sorting

Information on these items, traffic flow (always important given the high cost of transport) together with data specific to the site chosen will allow a layout to be planned.

Cold store layouts are invariably planned for future extension, so the information on which the initial layout is based will always be incomplete to some extent. Developing markets will dictate future development with regard to product characteristics and the need for special handling equipment and other investments that affect the product flow.

1.4 Mechanization

Bulk cold stores are always planned with some degree of mechanization in mind, from the simplest, basic fork-lift truck (FLT) operation to a fully automatic, computer-controlled, stacker crane operation. However, in all cases, a certain amount of manual handling will remain. This may involve unloading and loading of vehicles, stacking product on pallets, and occasionally sorting to mark or grade. In some cases, hand stacking will even be required in the cold rooms; for example, for long-term storage of carcass meat.

Routines and practices for manual handling are important for the economy of the operation and for work safety. It is important to acquire and have access to considerable expertise in this area. For these reasons, it may be advisable to consult specialists who will be better able to prevent costly errors in the planning stage.

1.5 Layout

In general, bulk cold stores are laid out for FLT operation which means that the chambers are built as a single storey with internal heights of 8–10 m. Handling costs and building costs, as well as increased demand for flexibility in use, have brought a universal development of large rooms, usually 15 000–20 000 m^3. Rooms are long and narrow, with doors only at one or both ends. Several rooms, side by side, face a common loading bank, which provides access to vehicles and to service and handling areas.

The working areas of the cold chambers and loading banks are usually at a height above the yard level, coinciding with the load level of lorries or railway wagons, in order to facilitate loading and unloading.

1.6 Storage methods

Block stacking

It is generally acknowledged that a system of block stacking of pallets in single-storey cold rooms achieves the best utilization of usable space within a large chamber. Whilst the layout of large chambers must be determined by the storage pattern to the lot size and turnover, it should be remembered that a deep stow will give a high net volume of actual product stored. However, deep stacking can mean much more taking down and re-stacking to locate particular lots and rotation units.

Although there is no recognized standard, a typical bulk chamber has a width of 25 m with two trucking aisles. Twelve rows of pallets are stored between the two truck aisles, and three rows are placed on the outer side of

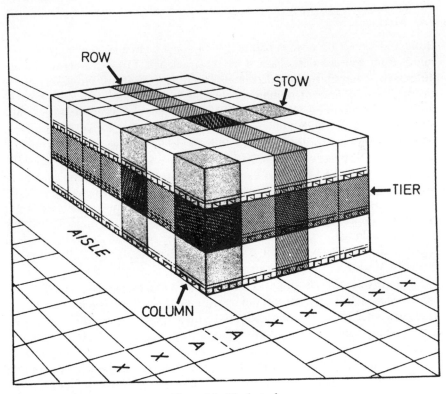

Figure 1.1 Block stack.

each trucking aisle. Access to the centre block of pallets is from both truck aisles, so that at no point will any stow be more than six pallets deep (these figures relate to a 1.0 m × 1.2 m pallet).

It is often possible also to use the front pallets of bulk stows for small lots including part pallets, provided that frequent access is not required to the bulk stack behind.

Terminology

The following definitions can be used as a guide when discussing the layout analysis of handling and storage problems and can be helpful in avoiding confusion:

- *Stow* Two or more columns behind one another (across gangway)
- *Row* Two or more columns beside one another (along gangway)
- *Column* Two or more pallets on top of one another
- *Tier* A layer of pallets in the storage block (column, stow, row or block)
- *Block* Several rows or stows adjacent to one another

The word 'stack' is used when it is not essential to distinguish between column, stow, and row.

Spaces

It is necessary that pallets in block stows have a space of not less than 0.5 m between the top of the columns of pallets and the ceiling of the cold room to allow for adequate air circulation. Space must also be allowed between the pallet columns in a stow of pallets, and between the stows themselves, of 50–150 mm, not only for air circulation but also to enable an individual stow of pallets to be drawn out of a block. This would, of course, be impossible if the stows were actually touching each other.

Pallet stacks must not, on any account, be allowed to touch outside walls. In most cold rooms, it is normal to install a kerb at the base of the wall to prevent this from happening. However, bad stacking can result in the upper pallets of a column touching the walls. Not only is this dangerous from a physical point of view, because the stack may fall, it also impedes air circulation. Pallets that are actually touching outer walls may suffer a rise in temperature.

Stacking discipline is all-important for the correct operation of palletized cold rooms. A pallet pattern should be painted on the floor; trucking aisle lines and stow lines must be strictly adhered to, and pallet columns should, at all times, be perpendicular. This is essential for safety as well as for efficient operation.

In instances where bulky products, such as carcass meat, are palletized and the product overhangs the pallet, it is important to leave sufficient space

Figure 1.2 In-store stows.

between pallet stows, so that interlocking does not take place if an individual stow has to be moved.

In a standard room layout, as described here, the actual space taken up by the product itself, not including pallets, assuming the cold room to be fully occupied, would result in a use of approximately 50% of the total storage space available.

Corner posts or frames

Many of the products placed in cold stores require some form of support to enable them to be palletized and stacked in the cold rooms. Such support is normally given by corner posts (also referred to as pallet posts) and converter frames. When designing cold stores, it is usual to dimension the chambers to allow pallets to be stacked four or five high. Most products are not strong enough in themselves to support the weight of such a column of pallets; they would collapse, causing a fall of pallets and damage to the product. Even when a column of pallets appears to be stable, the effect of weight on the bottom pallets can be cumulative. Over a period of time, the product on the bottom pallet may be compressed unevenly and cause the pallets above to tilt. If there is any doubt at all about the ability of the product to stand this stacking weight, then posts frames, or racks must be used.

It is important to remember also that if a product which is normally quite safe to store free-standing on pallets is received into store at a higher temperature than is customary, then the product must be placed in frames or

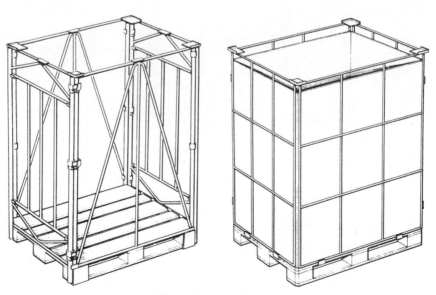

Figure 1.3 Special frames.

corner posts to allow for the temperature of the unit to equalize because this in itself can also result in a distortion of the stack.

Because of their shape, many products, such as carcass meat or irregularly shaped packages, cannot be placed on a single pallet without collapsing. Such products must be placed into special frames when they are being off-loaded from the vehicle delivering to the cold store. Since the customer has the same handling problem, it is sometimes possible to arrange for the product to be placed in these special frames at the time of processing at the customer's premises. In such cases, it is essential that each pallet is checked carefully for stability before being stacked in the cold rooms.

It should also be noted that it is often not the actual product itself but the material and type of packaging which influences the decision when and where to use corner posts or converter frames. Cartons can be placed in different patterns according to size: the pattern used should optimize stability and space utilization on the pallet.

Hand stacking of carcass meat

It is possible to stack carcass meat by hand in converted static racking frames. This method, when correctly organized, affords a better space utilization for difficult products. However, there are a number of important points to consider, of which two—hygiene and safety—come high on the list.

Substantial metal plates should be used between the beams on which the product is placed, and care should be taken not to overload the rack.

1.7 Chamber layouts

The initial design of a chamber layout can be made only on an estimate of the traffic and units to be handled. Often the picture which emerges is quite different from that originally painted. Customers may change their product distribution methods and consequently their handling and storage requirements. In such cases, the chamber layout should be reviewed. If large tonnages of the same product are to be stored for extended periods, as in the case of seasonal crops like peas, a significant increase in utilization can often be achieved by block stacking across one gangway for all or part of a chamber.

Examples of storage patterns for block stacking are shown in Figures 1.4 and 1.5.

The layout can be changed when changes in lot size and turnover occur. A third gangway with a 3×4 pallet layout module will, of course, make the pattern better suited for smaller lot sizes.

1.8 Racking systems

In most cases, pallet units stored in bulk chambers are associated with corner posts or converter frames of one kind or another. However, bulk chambers do,

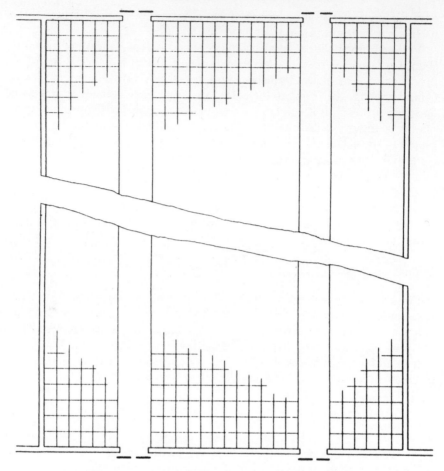

Figure 1.4 A standard layout with a 6 × 6 pallet module.

understandably, offer great flexibility with regard to varying products and space utilization. We must now look at alternative methods of storage using the various racking systems available.

Static racks

Racks consist of a steel framework which is built to support all pallets above ground level independently. This method gives direct access to every pallet in the system. The space occupied by the steel rack and the clearances required between pallets give far less efficient space utilization than block stacking. In addition, an aisle is required for every two-pallet row.

Static racking is installed for small lot sizes with high turnover or where

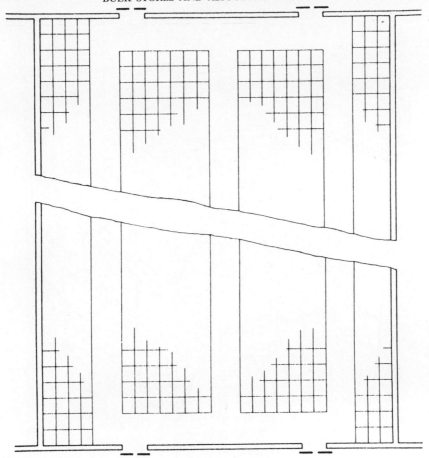

Figure 1.5 Room as in Figure 1.4 but with three gangways.

access to every pallet is important for other reasons. Static racking is common for break-up operations where, in most cases, the picking positions are at ground level, and positions above are used for spare pallets. Often only a limited area of a chamber is equipped with racks; in particular, static racks are never used for general-purpose storage.

Mobile racks

Racks for two-pallet rows are assembled to a carriage with wheels. These run on rails in the floor arranged to cross the length of the racks. Along the rails, there are so many racks from wall to wall of the chamber, that in one position only is there enough space between two racks to allow an FLT to operate. By

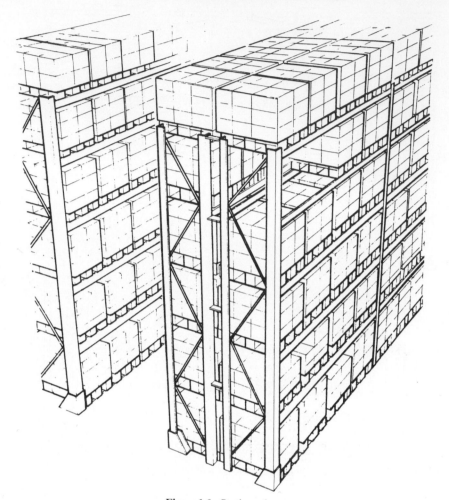

Figure 1.6 Static racks.

moving some of the racks, this space can be opened up between any two racks to give direct access to all pallets in the system. Each rack is motorized and connected to a control system that allows the FLT to open up the desired position with a push button or possibly a remote control, either from the truck itself or from some external control point. Mobile racks offer better space utilization than static racks, but access is not quite as quick. This type of racking is ideal for small lots with a limited turnover, where space utilization is important.

Recently, developments in the field of mobile racking have proved that this type of installation can be used very successfully with turret trucks which,

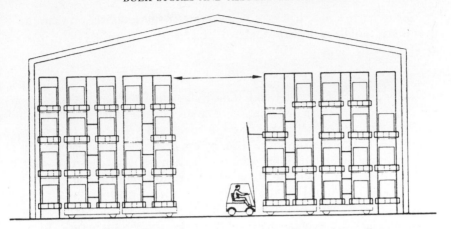

Figure 1.7 Mobile rack installation with static racks along walls.

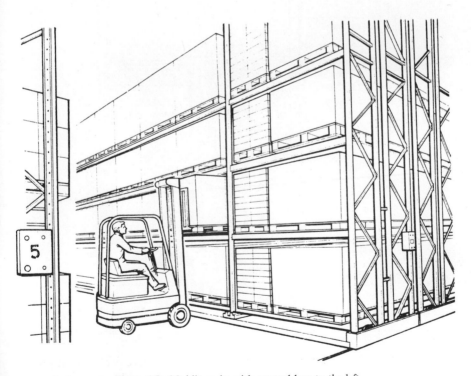

Figure 1.8 Mobile racks with control box to the left.

operating as narrow aisle trucks, offer even better space utilization than the conventional mobile rack with standard FLT.

Drive-in racks

Drive-in (or drive-through) racks are static racks with several pallet positions behind one another as seen from the aisle. The steel structure allows an FLT to enter empty racks because there are no cross-beams; see Figure 1.3. The pallets are supported only on two opposite edges by longitudinal beams for each tier.

Drive-in racks offer individual support of each pallet and better space utilization than static racks. The stock is accessible on the LIFO (last in, first out) principle which limits the use of this system. It may be used as a buffer store in break-up operations or as a terminal store for loads prepared for outloading. The location of the pallets in the racks is important in order to get good utilization.

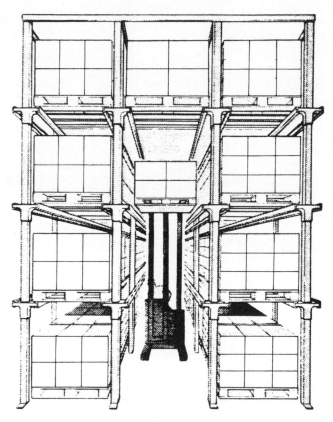

Figure 1.9 Drive-in racks.

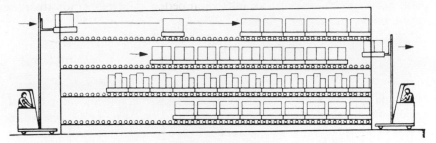

Figure 1.10 Live storage.

The most obvious advantage of the drive-in rack system is that it can easily be used in older and less adaptable bulk chambers and also dispenses with the use of corner posts and metal converter frames.

Live storage

A live storage block consists of several lanes of roller conveyors arranged side by side and in tiers above one another. The conveyors are slightly inclined, so pallets loaded at the upper end will advance through the block by gravity. A braking arrangement ensures that they do not move too fast. At the lower end, the pallets are stopped and queue up in the lane until the first pallet is removed.

Live storage is a system for big lots and high turnover. The outloading capacity is high and the FIFO (first in, first out) rotation is easily maintained. The system needs careful calculation because the utilization of pallet positions is, in practice, not easy. Investment per pallet position is also high. Live storage systems can also be used successfully with break-up operations where the system itself is positioned at high level above the order-picking aisles.

Rack entry module

A rack entry module (REM) system consists of a steel structure similar to drive-in racks. The pallets are moved by a self-propelled carrier module which travels along each lane. The carrier module is shifted horizontally and vertically from lane to lane by a transfer lift moving along the one face of the block.

The REM system may be manually operated, semi-automatic or computer-controlled. It is a compact storage system with access to more lots than the conventional block storage. The system can be installed in high-rise buildings but gives much less flexibility than the stacker crane system. It is also possible to use the racking system to support the building, if the system is designed to be a permanent installation when a new chamber is being constructed. However, this decision must be taken very carefully as the REM system does have some disadvantages, as does the live storage system.

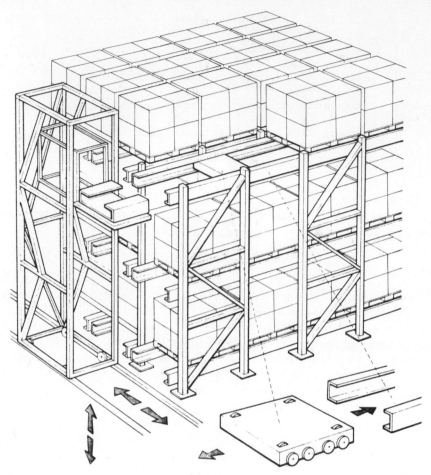

Figure 1.11   REM system.

Stacker crane storage

A high-rise store is one that stores goods at a greater height than that which can be handled by FLTs. The term is restricted to layouts where the pallets are supported on steel or concrete structures, which can be 30 m or more high. In addition to pallet loads, they can support stacker frames, external walls, mezzanine floors, etc.; in other words, this installation can also be the supporting structure of the warehouse, as described for the REM system installation above. A variety of handling equipment is available. The system can be controlled manually, semi-automaticaly, or fully automatically. All stacker crane installations are designed specifically for a particular situation.

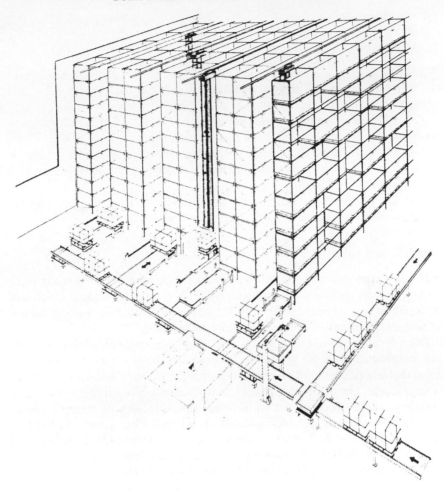

Figure 1.12 Stacker crane store.

They can provide highly efficient and cost-effective automation of handling and storage of large stocks with a fast turnover.

To be effective, stacker crane installations must be planned in great detail to perform the functions required. Initial costs per pallet position are higher than with pallet racking, and flexibility is more limited. Preventive maintenance and servicing programmes are of particular importance for a reliable operation. A stacker crane installation will undoubtedly come into its own when the turnover of large quantities of product exceeds 10 times per annum. It is also important to have a steady product flow for such a system, which can be programmed over a 24-hour period.

1.9 Stacking material

Standard pallets

Within the frozen food industry, and the food industry in general throughout the world, two standard sizes of pallets are in normal use: 0.8 m × 1.2 m and 1.0 m × 1.2 m. Both have a thickness of 15 cm.

The pallet most commonly used today is the 1.0 m × 1.2 m, which seems to have advantages in all links of the distribution chain.

The UK frozen food industry has a recommended standard for a 1.0 m × 1.2 m pallet. It is a four-way entry unit; where possible, it should be constructed of soft wood.

The use of the 0.8 × 1.2 m pallet in the UK is very limited, and no recommended specification is currently available for it.

Non-standard pallets

All other types and sizes of pallet are non-standard. Within the cold store industry, where there is a steady demand for the storage of product which will not easily fit on a standard pallet without an unacceptable loss of space, the use of non-standard pallets can occasionally be justified.

Disposable pallets are used only where the distances involved in transporting the product from the producing country to the consumer market are so great that any possiblity of pallet return or participation in pallet pool schemes is impractical. The disposable pallet usually consists of a number of blocks of very low-quality wood or carboard fixed to a thin base; its strength comes mainly from the way in which cartons on the pallet are strapped together. When receiving product on disposable pallets, it is essential in the interests of safety that the product is transferred to a standard pallet, if necessary with a convertor frame, before it is stacked inside the cold chamber.

Quality standards

The pallet board is undoubtedly the most important piece of equipment used in warehouse systems and, as such, must be maintained with the highest degree of care. A poor-quality or damaged pallet can prove extremely costly not only in terms of damaged product but also occasionally in causing serious injury or even the loss of life.

Pallet posts

Pallet posts or corner posts are designed to give stability to a pallet and also to take the weight of pallets stacked above.

Individual corner posts are secured to the pallet unit by means of steel banding using either fixed top and bottom timber frames or a number of

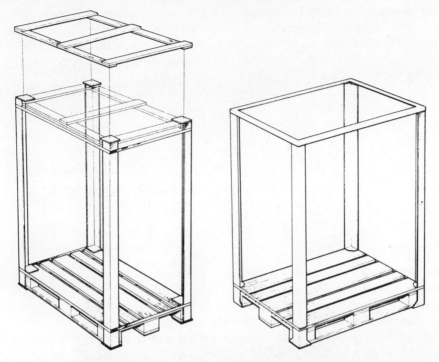

Figure 1.13 Corner posts with wooden and steel top frames.

timber spacers positioned inside the angle of each post. In the case of carton packaging, when the product exactly fits the pallet without any gaps in the pallet layer, both the carton and the product may be strong enough to withstand the pressure of the banding. Then posts can be fixed directly to the pallet unit.

If, however, there is any doubt as to the strength of products, or the product itself is unstable, then wooden slat stabilizers or metal bars must be used to fix the tops of the corner posts to each other. It is of course essential that banding is sufficient and tight before stacking of such pallets commences.

Where possible, pallet posts should be galvanized; painted posts are very prone to rust, especially as posts are subject to impact damage of one sort or another in the course of normal handling.

Pallet frames

Standard pallet frames are normally sized to fit the 1.0 m × 1.2 m pallet. They are made in a multitude of designs by many manufacturers. When handling heavy and unstable products like carcasses in pallet loads, rigid frames must be used because conventional corner posts do not provide acceptable safety for the product or the unit. In principle, pallet frames are made to be collapsable but in practice frames, once erected, are often used as semi-permanent solid units. This fact often leads to a situation where the wooden pallet becomes badly damaged yet continues to be used because the frame holds the wooden pallet together—a practice which is extremely dangerous.

Pallet frames must be regularly inspected, both for physical damage and also for corrosion (especially if frames are stored in the open, as they often are). Only frames in good condition should be taken into use in the cold chambers for any variety of product.

The cost of pallet frames and converters is relatively high compared with that of pallet posts. Converters and frames should be used only if there is a fairly high turnover, if products are awkwardly shaped, or if a large number of part pallets are likely to be involved during the course of the storage period.

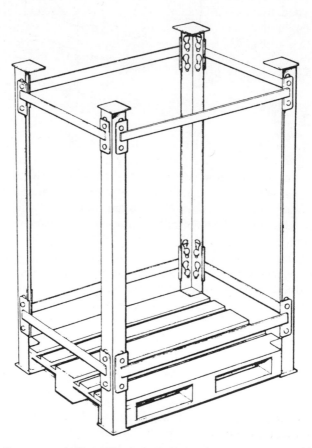

Figure 1.14 Cornner posts with crossbars attached in keyhole slots to give stability close to that of a rigid frame.

Strapping material

The usual type of strapping material used in securing pallet posts is steel banding. This requires the use of a banding machine. The degree of tension of the banding must be correct; if it is too tight, the banding will contract in the low temperature of the cold chamber and may snap, and if it is too loose, the pallet stacks will be unstable.

It is dangerous to use steel-banded pallet units more than once; new banding should be applied after each use.

When pallet post sets are dismantled, care must be taken to gather up the discarded banding and remove it from operating areas of the cold store. Steel banding is not only unsightly but can cause serious damage to wheels of FLTs and pedestrian trucks, and even injury to personnel.

Automatic and semi-automatic banding machines are available on the market. A fixed strapping station with sophisticated arrangements can always be justified where there are a large number of units to assemble at any time.

Other strapping methods, such as stretch-wrapping or shrink-wrapping of products with plastic film, are used for some products. This types of strapping is necessary for pallets without corner posts when these are stored in racks, and is becoming increasingly popular.

Spacers for freezing

As part of the stacking material back-up for any bulk cold store, it is quite usual to find fresh product being frozen either in low-temperature cold store chambers or through a conventional blast freezing tunnel. When cartoned products are placed in the freezing tunnel, an optimum air circulation must be achieved to ensure rapid and even freezing of products.

In order to achieve this, it is necessary to place spacers between the layers of cartons on the pallet. Figure 1.15 shows a typical pallet set-up with such spacers interleaved.

In freezing tunnels, it is also important that space is left between pallets placed next to each other in the direction of the air flow in order to ensure an even freezing pattern.

Freezing frames

There are various types of specialized frame designed for use in blast freezing tunnels. Common types in use are for the hanging of meat carcasses for freezing, and frame racks for cartoned products.

The size of the frame may vary according to the actual product being frozen; it should be noted that the frames should be capable of being stacked in the cold room, since either the workload of the loading bay staff may be such that it is not convenient to discharge the product on to pallets for final stacking at

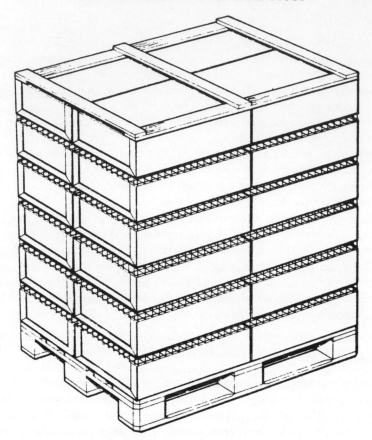

Figure 1.15   Pallet of meat cartons with spacers.

the time the frames are removed from the tunnel, or the storage time may be too short to make such a transfer economically worthwhile.

Since carcass meat for freezing is delivered for freezing to the cold store in a hanging position, the freezing frame should be so adapted that the hanging bars can be connected to the meat rails system, at the point of delivery from the transport vehicle.

Palletainers

It is current practice throughout the world to place vegetables and some fruits into bulk packing units directly from the blast freezer, and then to repack into appropriate retail or catering packs as market demands require. One common type of bulk pack is the palletainer.

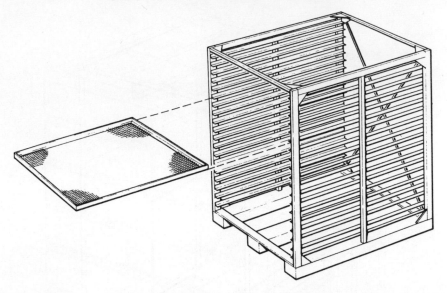

Figure 1.16 Freezing rack for packaged products.

Palletainers can be of varying sizes, according to the product stored, but the most usual is that which is designed to hold 1 tonne of peas.

Product inside a palletainer should always be covered to prevent dehydration and to give the same level of protection as for a product packed for retail sale. This is important to reduce drying out which in turn increases the weight loss of the product if it is held in store over a long period of time.

It is important to ensure that the palletainer is assembled with a great degree of care and accuracy; a badly constructed palletainer unit can endanger the whole stack if one unit is slightly off the vertical.

Angle steel corner posts are generally used as the supportive unit for palletainers, although timber corner posts forming an integral part of the card box are also used. In either case accuracy in assembly is all-important.

Control of stacking materials

The stacking materials form a significant part of a cold store investment. Such an investment must be protected by rigid control and good maintenance practices. Clean and well-maintained stacking materials are important to the safety and protection of the goods stored in any bulk cold store.

Stacking materials not in use should be marshalled in a defined area, allowing supervision to inspect and check quantities at regular intervals. All damaged pallets must be withdrawn from the system and repaired if possible. Where corner posts and metal converters are damaged, they should be scrapped. Where deterioration has taken place through corrosion, these

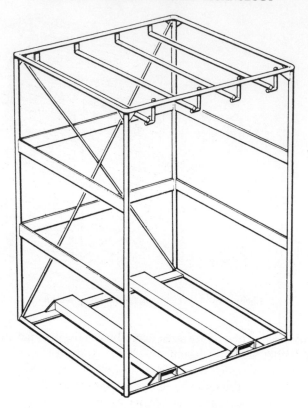

Figure 1.17   Frame for freezing of hanging carcass meat.

materials should be accumulated separately until they can been cleaned (usually with a shot-blasting machine) and then repainted. The quality of the finished paint surface is important, as only certain paints are acceptable in the presence of foodstuffs. In the case of blast freezing frames, where the equipment is under even greater stress because of temperature variations, a galvanized finish is recommended and in certain special circumstances, stainless steel is the only metal acceptable.

It is normal for bulk cold stores to receive goods on pallets, and to replace the pallet received with an exchange pallet. It is of the utmost importance to ensure that only an acceptable standard of pallet is received into the store. Damaged pallets must not be accepted for storage, as this could result in the collapse of a stow and possible injury to personnel. The product must be transferred to an acceptable unit and the customer advised and charged with this operation. It is possible to use a machine called an invertor to replace damaged pallets, but of course this equipment is only generally useful where uniform carton products are stowed on the pallet board.

Figure 1.18 Palletainer of cardboard, wooden slats and corner posts.

Inloading documents should be endorsed when defective stacking materials are shipped into the cold store. It is most important that an inventory is kept at all times of stacking materials in circulation, as in a large bulk cold store, stacking materials amount to an enormous cost item.

1.10 Fork lift truck handling equipment

Pedestrian trucks

A truck operated by a man standing on or walking with the truck is usually referred to as a pedestrian truck. It is capable of lifting a pallet purely for transportation purposes. The lifting height is generally limited to 10–30 cm. Small roller wheels extend from the forks of the truck to take the weight of the pallet and to give stability.

Pedestrian trucks may be either electrically or manually operated with a hydraulic pump lift. Electrically-powered appliances are most commonly used in bulk cold stores.

One of the principal jobs of the pedestrian truck is to unload pallets from insulated or refrigerated vehicles via a bridge plate to the loading bay. It is important that the type of truck chosen is capable of negotiating the relatively steep slope of the bridge plate. It is also important that the main drive wheel and roller wheels attached to the forks are wide enough to ride over the floor corrugations which are an integral part of the design of modern refrigerated vehicles and containers.

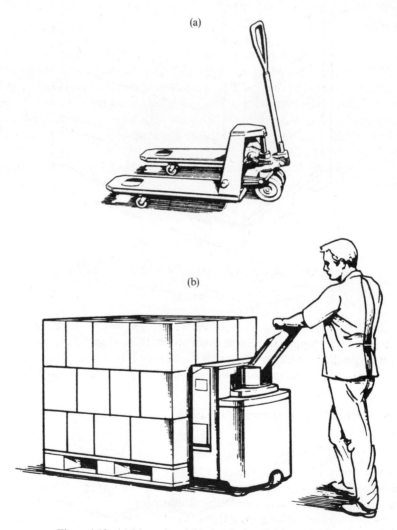

Figure 1.19 (a) Manual and (b) electrical pedestrian trucks.

The operator must always proceed with the pedestrian truck whilst it is in transit. From an operational safety point of view, the pedestrian truck, be it stand-on or conventional, should never be moved with the pallet in the lead position. The exception to this rule is when placing the pallet in position in the transport vehicle on a weighing platform or in other similar areas which require precise manoeuvring.

Ride-on trucks

These trucks usually travel considerably faster than the conventional pedestrian truck. They are often used when relatively long distances have to be covered, for instance in moving product from the bulk cold store to the inspection or repacking area. However, for vehicle loading and unloading, the conventional pedestrian truck is more manoeuvrable and more practical.

Manual hydraulic pump pedestrian trucks can handle the same load as electrical trucks, but obviously more labour is involved and the operation is much slower. However, manual trucks are much lighter and easier to handle. They can be of considerable use with road vehicles delivering palletized loads to points where FLTs or conveyors are not available. Manual trucks are also used for such tasks as moving engineering spares and stores where the utilization would not justify the investment of a electrically-driven pedestrian truck.

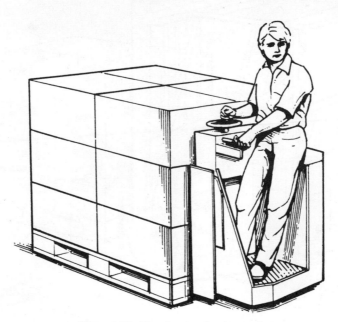

Figure 1.20 'Stand-on' pedestrian truck.

Under no circumstances should pedestrian trucks of any sort be used for yard work, since the relatively rough surface of the tarmacadam or concrete will quickly damage the roller wheels in the forks and stabilizing wheels on the truck itself.

Counterbalanced trucks

A counterbalanced FLT is a truck where the pallet is both carried and lifted outside the overall wheel base of the truck. The rear of the truck is weighted to counterbalance the pallet weight at full elevation. Such trucks operate with fixed masts. To place the pallet on a stack or vehicle, the mast itself is tilted and the truck is slowly driven forward.

Counterbalanced FLTs may be powered by batteries, by diesel, or by gas-operated internal combustion engines. Internal combustion engine trucks are used for outside work in yards, etc., and battery-operated trucks are used in cold rooms and on loading bays.

From a safety point of view, it should be noted that all FLTs should be fitted with fire extinguishers and lights.

Lights should be placed in such as position that they ensure reasonable visibility when placing top pallets in stacks in the relatively dark areas away

Figure 1.21 Counterbalanced FLT.

from the trucking aisles where the lights of the cold room are generally positioned.

For operating in block stacks, a side shift of the fork lifts is a usual attachment to the truck. Note that the side shift reduces the lifting capacity which is important when selecting the particular truck to be used.

Service and availability of spare parts is also of primary importance when choosing an FLT.

Reach trucks

The principle of a reach truck is that the pallet and mast is carried within the wheelbase of the truck, allowing the use of narrow gangways. To place the pallet, either the mast moves forward or the forks are projected by a pantograph mechanism.

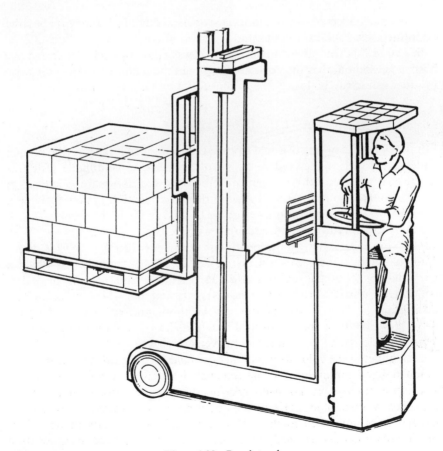

Figure 1.22 Reach truck.

In comparison with the counterbalance truck, the design of the reach truck allows narrow gangways and better stability which means higher lifting capacity. On the other hand, a counterbalance truck has the advantage in horizontal movement because it is faster and move manoeuvrable.

The wheels of the reach truck are much smaller than those of the counterbalance truck. Consequently, reach trucks should never be used for yard work because of the relatively rough surfaces involved.

Like counterbalance trucks, they require lights and fire extinguishers.

Considerable research and development has taken place in recent years with regard to reach trucks, and they are now more reliable and more sophisticated than the earlier models. It is now possible, for example, to leave a reach truck inside a low-temperature cold room for the whole of the working shift; and the truck comes out of the room only for battery changing.

1.11 Administrative routines

The service a bulk cold store offers its customers is directly dependent upon the administrative back-up routine that it has in operation.

In a large bulk cold store with its many stows of pallets, stacked four and five high, it is essential that the documentation is simple and accurate, otherwise product can easily be lost in the mountain of units in the chambers.

Location system

For the efficient working of a cold store, it is necessary that the position of each and every pallet in the cold room should be clearly identifiable. For this purpose, a pallet location system must be operated and maintained on an almost hour-to-hour control. Not only does the location system provide a vital link in the store operation, it can also be used as a counter-check for stock records in the event of any discrepancies arising.

The basic information for the location system is provided for the FLT driver who completes a location sheet giving the actual stow number in which a particular pallet is stacked and also the position of the column of pallets concerned within the stow. It is, of course, essential that when any restacking takes place in the cold rooms, the new locations are reported by the FLT drivers. Normally, a special form of location sheet is used for this purpose, generally referred to as an internal movement sheet.

Each pallet in the cold store bears its own individual card, each card being numbered and containing information such as customer, type of product, lot or rotation number, etc. To ensure product safety, pallet cards are fixed only to the centre wooden block of the pallet; under no circumstances should cards be fixed to the product itself. Each card has a tear-off part giving the number. This portion of the card is attached to the location sheet and passed to the location clerk.

Figure 1.23 Pallet card.

The actual location information is usually kept in a card system which should be able to provide the following information:

- The situation of each pallet stow in the cold rooms showing details of product stored in the stow.
- By cross-reference, the identification of location by customer and lot number.
- The location of a pallet from a pallet card number.
- A separate record should be kept to check that each pallet card issued is actually used and that no card have been inadvertently left in the system after a particular unit or lot has been delivered.

It is important to ensure that all pallet cards are removed from pallets when outloading takes place. It is usual for such cards to be returned with the out-loading documentation sheet so that they may be removed from the system.

Sometimes the location sheet is used to determine where pallets are to be placed when inloading and in these cases, the FLT driver receives pre-marked location sheets.

Modern technology is now moving rapidly towards computerization and radio control between the office and the FLT driver, which should mean that the accuracy of locating unit pallets will become almost foolproof.

Stock records

The purpose of stock records is to maintain accurate and up-to-date information about customers' products held in the cold store. Accurate records also provide the base from which to determine the important question of charges in respect of services carried out.

Today, the tendency is for records to be computerized. The older method, still often used in smaller cold stores, is some form of card index system. Whichever method is used, it is essential that information on product movement over the loading bank is entered into the stock record system continuously and as quickly as possible.

It should be remembered that very often the customer's own records are not up-to-date or correct. If the cold store is not completely in command, almost on an hour-to-hour basis, mistakes will occur such as wrong product being delivered, and, although the fault may lie initially with the customer, the cold store may be blamed for not informing the customer of his error at the time his instructions were given.

The basis of all information for stock records is the checking sheet and/or delivery document compiled during the process of receiving or despatching product. The greatest care must be taken both in the actual checking and in ensuring that the checking document is processed without delay.

The normal flow of checking documents is via the chargehand or foreman to the despatch office. Before any delivery instructions are given, it is often advisable to check the stock record for availability of the product lot concerned and to make some form of temporary notation that the delivery is in progress, thus avoiding any possibility of the customer over-ordering on any particular line.

Stock control

It is generally recognized that in the warehousing industry throughout the world, the term 'stock control' means not only that stocks held on behalf of the customer are properly recorded but also that such stocks are rotated on the basis of 'first in, first out', the so-called FIFO principle. However, it is considered advisable that the customer should give specific instructions on how he wishes such a system to be operated. This principle is particularly important in the case of chill storage where product life is, almost always, rather limited.

It must be borne in mind, however, that in the case of small lots, extra handling may be involved in observing the FIFO principle. Indeed, where bulk stows with sizeable tonnages are involved, it may be either uneconomical or impractical to operate. This is particularly the case, for example, where large stows of hand-stacked carcass meat or bulk stacks of vegetables are involved.

It must be appreciated that it is virtually impossible to take a complete physical stock check of a cold store holding at any one time, or to check stock for any one large customer whose goods move rapidly. It is also impossible to take an accurate stock check of hand-stacked products. Customers' requests for physical stock checks to be taken at one particular time, for example at the end of a financial year, should generally be resisted, or additional charges made for this service.

It must be appreciated as a medical fact that temperatures as low as $-30°C$ have an adverse effect on the powers of human concentration after a period of exposure, making stocktaking of any sizeable quantity of products quite difficult. Stocktaking should, therefore, be carried out in short working periods throughout the year, whenever the pressure of work is such that the loading bank staff are not fully utilized. Staff carrying out such checks should never be informed of the booked stocks beforehand.

Each pallet row should be checked for the actual product stored in that row, and the result compared with the pallet location system. It is, of course, also necessary for stock records and location systems to be regularly spot-checked to ensure that they are in agreement.

It cannot be emphasized too clearly that an efficient stock control is one that is not only accurate but is up-to-the-minute as far as feasibly possible. Stock records which are 100% accurate but a couple of days behind the actual operation, are inadequate for the demands of a modern cold store operation. The tendency today is for stock records to be computerized, but unless the appropriate information is fed into the computer quickly and accurately, the benefits of this will not be fully realized. It is also important that the person entering stock information into the computer has some knowledge of the cold store operation itself, in order to avoid placing any obvious mistakes on the computer record.

1.12 Additional services

Additional services associated with bulk cold stores can be many and varied, depending upon the customers' needs and the cold storage operator's desire to satisfy those needs in order to secure his business.

There are too many services to detail in this chapter, but some of the most popular and widely offered are worth mentioning.

Packaging

Packaging of frozen product is mainly associated with frozen vegetables, which undoubtedly command a large volume of the space in most cold stores. Fruit is another product in this category, and the whole range of fruit and vegetables can now be handled through sophisticated packaging machines which are totally automatic and are technically sufficiently advanced to produce a range of accurate packs that meet the legal requirements of the Weights and Measures Officers and of the local Hygiene Inspectors.

This equipment, which needs to be located in specially constructed premises, in order to meet a high standard of hygiene, is extremely expensive. The cold store operator must have an assured and guaranteed contract before speculating with such high-valued equipment. Once having taken this step, it is then necessary to support the venture with the level of skills needed by the supervising people, both operational and engineering, to ensure maximum utilization of the equipment.

In addition to vegetable packing, of course, certain meat packing is undertaken, particularly in the field of carcass lamb which is cut up into various portions, weighed, and packed for retail consumption. Again, the premises in which this work is carried out have to meet a very high standard in order to satisfy the local veterinary inspectors.

Freezing of meat

Industrial freezing of meat has been used for about a century. It was first used to facilitate transport from overseas countries to markets in Europe, but cold storage was also used to take care of the seasonal surplus of meat. Microbiological deterioration was stopped by cold storage, but otherwise quality was impaired because of poor techniques in freezing and storage. Frozen meat therefore acquired a bad reputation in the early days: an opinion which has, to some extent, remained. Many consumers and also meat trade people look upon frozen meat with some suspicion, but meat properly frozen with modern techniques and stored under appropriate conditions is, as far as the overall quality is concerned, almost indistinguishable from chilled meat.

When a very high water-binding capacity is required for some meat products, the meat is theoretically best frozen immediately after slaughter when the pH is high. It is, of course necessary to avoid thaw rigor. However, it is impossible to freeze a carcass or a quarter-carcass fast enough to achieve this objective. The meat must therefore first be cut and then frozen by a rapid method.

Carcass meat is normally frozen after complete chilling. Such meat must be deboned before use, and this requires thawing. If the freezing is performed on the third day after slaughter, the water-binding capacity is very low and the subsequent drip when the meat is thawed is high. The better cuts of meat are not sufficiently aged, and therefore will probably be tough. Such meat is therefore used for process meat products.

From a technical point of view, there are few problems in the freezing of whole carcasses. Blast freezing is the only practical method for carcass freezing. If a good air circulation between the carcasses is provided, the freezing will be fast enough to meet the quality criteria. The freezing rate is of course rather slow because of the large cross-sections, but the quality of frozen meat has been shown to be dependent upon the freezing rate only to a limited extent. It is, however, very important that the surface layer is frozen fast enough to prevent microbiological growth and this is normally the case. Meat from healthy animals is almost free of microorganisms in the inner parts and therefore microbiological problems do not play a significant role in carcass freezing, even if the freezing rate is rather low.

When meat is cut, the surfaces are infected. Since the ratio of surface to volume increases with the degree of cutting, the microbiological problems will increase. When such products are packed into cartons, it is therefore very important, from a hygienic point of view, that the freezing is performed as soon as possible and as fast as possible. Even with good air circulation between the cartons, it takes longer than is usually believed to reach the freezing point in the centre of the carton and therefore some growth of microorganisms is unavoidable.

The best way of freezing products such as meat cuts and meat products is in-line freezing before packaging. When meat cartons are frozen, the freezing must be carried out in a proper freezing tunnel. Instructions for the use of freezing racks and other equipment as well as stacking patterns in the tunnel must be followed in order to avoid unnecessary quality deterioration and to minimize the risk of future claims which otherwise might be caused by too heavy a microbiological growth during freezing.

Storage and thawing of meat

The storage conditions for frozen meat are most important, and it is essential that during storage meat is protected as far as possible from losses due to evaporation. There are various methods of preventing excessive drying or dehydration. A sufficiently low and constant temperature gives good protection, since the vapour pressure over ice decreases very rapidly with decreasing temperature. A storage temperature as low as $-30°C$ is therefore always better than a higher temperature.

Thawing is a very important but often neglected process. During thawing the same amount of heat must be supplied which was extracted during the freezer process, and the heat must be transferred through a deeper and deeper layer of unfrozen (thawed) meat to the inner parts.

Since unfrozen meat conducts heat less well than frozen, thawing takes longer than freezing. The surface layer is kept at a temperature which favours microbiological growth, and so there is a risk of microbiological deterioration. It is therefore necessary that the thawing of meat is performed under controlled conditions and in a hygienic manner, with precautions being taken to avoid excessive microbiological growth.

1.13 Summary

The image of the bulk cold store has changed considerably in the last 30 years. The demands on the bulk cold store today are such that it must be adaptable to the customers' ever-changing requirements.

Management and efficiency are the key to any successful operation, and these, combined with innovation, will always play a major part in any bulk cold store operation.

Initial planning is important if the development is to be capable of meeting the demands that may be placed on the operation over the years.

Acknowledgement

The author acknowledges the assistance of Alf Rasmussan, General Manager of the Technical Centre, Frigoscandia AB, Helsingborg, Sweden, who reviewed the draft manuscript.

2 Distribution depots and vehicles

B. BARNES

2.1 Introduction

The creation and growth of the modern frozen foods industry has taken place since 1945 and has been linked to sociological changes arising from World War II. These changes have included the redevelopment of cities, increased urbanization of all towns, reduced use of gardens and allotments and—perhaps most importantly—continuation in the employment of women away from the home. A new concept of 'convenience' was introduced by the food industry: the term *convenience foods*, meaning selected prepared, uncooked foodstuffs, initially simple products such as potato chips, peas, beans, etc., progressively came to mean specially created items, such as fish fingers, all preserved by deep freezing. The current meaning has now been further extended to included fully prepared frozen meals, ready to eat after a simple cooking process.

Of necessity, development of the frozen food industry was inextricably linked with home-ownership of domestic deep-freeze cabinets and fridge–freezer combinations. Initially the industry promoted the development of suitable equipment, and when this stage was achieved, large retailers joined in with the provision of facilities for the purchase of domestic equipment at reasonable prices. This set the stage for the frozen food industry to participate fully in the retailing revolution of the last ten years.

2.2 Historical background—before 1939

Before World War II only two industries in the UK used below-ambient distribution on any regular basis; fish and ice-cream. It is of course true that frozen meat from Argentina and New Zealand was freely available in butchers' shops throughout the UK, but the emphasis was on the product reaching the UK in good condition in refrigerated ships, rather than controlled temperature distribution. Larger butchers had a cold room, but this served a tempering role for frozen carcasses as much as a storage role.

Fish was gutted at sea, boxed with ice, and was kept iced until it arrived at the fishmongers. This was its fundamental protection. According to its route it might well have spent some time in chill rooms at the dockside or in insulated containers forming special fish trains from the major ports such as Hull, Grimsby, and Fleetwood. Ice would be replenished at major transfer points such as the dockside to the train. Major towns and cities had ice producing and retailing companies, and larger fishmonger shop chains had chill refrigerated storage at the shops. Importantly, the customer expected to see fish in fragmentary ice on the slab and was conditioned to eat fish that could have been up to two weeks in ice since being caught. Only in the locality of the smaller fishing ports was 'fresh' fish truly fresh.

Ice-cream was frequently sold by small local businesses direct from the manufacturing unit via insulated containers. In other words, the point of sale was also the point of production (like post-war 'soft ice-cream' mobile retailers). Major producers with a distribution network used solid carbon dioxide blocks to achieve temperature control. Ice-cream could be distributed in three basic ways:

- By a street retailer working out of an insulated container which carried a block of solid carbon dioxide with the ice-cream
- By despatch of insulated boxes to customers such as shops or larger restaurants (this method persisted for transport to the Western Isles from Glasgow until the 1970s)
- Despatch from the factory to a depot in insulated vehicles with solid carbon dioxide, transfer to large top-loading insulated chests, again using solid carbon dioxide, and then to street retailers or shops as before

Figure 2.1   A T. Wall's & Sons Ltd ice-cream delivery vehicle (*c.* 1920s).

Major ice-cream manufacturers had their own production plant for solid carbon dioxide block production.

Figure 2.1 shows an ice-cream delivery vehicle from around 1926. This vehicle delivered the insulated boxes described earlier.

2.3 Post-war growth of the UK industry, 1945–1980

The first segment of the industry to prosper was ice-cream, perhaps because company organizations existed and the two brand leaders, Wall's and Lyons, had sound financial backing. Rationing of commodities gradually ceased and preferred ingredients again became available, and new patterns of distribution quickly developed.

Small depots were set up in each town. These depots were supplied by solid carbon dioxide refrigerated insulated vehicles from the manufacturing factories. Each depot had a cold store, perhaps no more than $15\,m^3$, most likely refrigerated by coils served by a single-stage plant using R12. The product was unloaded by hand into the cold store, using elementary gravity roller conveyors where possible, and put to stack in bays. This was the heyday of van selling, so each van would be filled with a suitable selection of the product and sent on its route with the driver-salesman, or merchandiser as he was sometimes called, trained to sell-in, take orders, deliver there and then, and finally take money from cash customers. All this was assisted, of course, by the presence in the shop of a so-called 'conservator' provided by the supplying manufacturer.

In the early post-war days the pre-war pattern of distribution continued, a pattern firmly based on solid carbon dioxide and insulated boxes or, later, insulated box vans. Around 1960 a profound change took place, and refrigerated vehicles were developed. The vehicles used for shop delivery were refrigerated by means of eutectic plates which provided ten hours' 'hold over' during the day. At night they were recharged by a compressor carried on the vehicle and plugged into a mains electrical supply. For single-shift operation this system was very reliable and was in almost universal use. Vehicles were quite small and carried between one and two tons of product.

This resurgence of ice-cream sales after the war was catered for by existing companies who had held on to manufacturing facilities throughout the war. The fledgling frozen food business, however, started virtually from scratch. It produced a range of relatively simple products, such as vegetables and fish, and it sold through different outlets—the grocers and provision merchants rather than the confectioners, tobacconists, and newsagents of the ice-cream trade. Most importantly, from the start (and to the present day) it encouraged the development of display refrigeration in shops, but did not enter into the provision of shop cabinets as the ice-cream business did. Few home deep freezers were available at this time, and both ice-cream and frozen food was in practical terms sold for same-day consumption. Those homes with refrigerators had at best, only a compartment or a tray where freezing or just sub-freezing conditions could be maintained. In the absence of consistent deep freeze facilities in the home, the practice of marking frozen food with stars to indicate the desirable maximum period of storage was introduced.

In the late 1950s significant growth of the frozen food industry was taking place and distribution had to keep pace. In general the distribution effort remained directly or indirectly with the manufacturers as, at this time, only they saw the need for investment in distribution. Cold stores were built on a large scale, palletized goods from the factories were stored up to three pallets high, either on posts or in early racking system. This in turn required the development of battery-operated FLTs capable of lengthy periods of work at low temperature. It became usual to use blown air from evaporator coils to refrigerate the rooms, instead of static coils.

A feature of the period, not thought unusual at the time, was the very high rate of stock turnover in the early stores. In part this was because the concept of distributed stock had not taken hold, and in part the rapid growth in this period of the whole industry meant that the cold stores were usually undersized. This feature lead to very substantial installations of plant. In the summer a store of this type could be turning over its stock three or four times per week! The significant item here was stock turnover, and plant capacities were frequently six to ten times as high, per unit volume, as the bulk stores of the day.

The use of wooden pallets, at that time normally the 1 m × 2 m four-way entry pallets, grew as the use of battery-operated pallet trucks grew. Low-temperature materials handling became a new segment of the established materials handling industry.

This maturation of the industry continued at an accelerating rate during the 1960s, with more and more investment taking place in new depots, new cold stores and, most noticeably to the public, new vehicles in bright liveries in all the High Streets (Figure 2.2).

Larger depots now tended to serve an area, not just one town. Street retailing of ice-cream was franchised out to sales companies who frequently operated with self-employed drivers. This was the second step in the movement of the point of sale from the manufacturer's orbit to specialist retailers. (The first was cabinets in shops.) A consequence of this was that individual deliveries became larger and concepts such as 'drop value' and 'drop size' became important. Perhaps it could be said that at this stage a fledgling frozen product distribution industry was formed.

Shop delivery vehicles were now carrying up to $2\frac{1}{2}$ tonnes of product. They were also carrying nearly a tonne of eutectic plate and refrigeration plant. Nothing more reliable had been developed and despite its penalties in weight and general clumsiness, the eutectic plate remained in almost universal use.

Towards the end of this decade another revolution began, a revolution that was to have the most lasting and profound effect. Supermarkets became

Figure 2.2 A T. Wall's & Sons Ltd ice-cream delivery vehicle (*c*. 1960s).

significant. This of course meant the end of van selling, and the early experiments in selling by telephone, very quickly named 'Teleselling', were so successful that the method became universal with the larger suppliers, even when it was associated with a manual docket method of recording sales and subsequent paperwork. Very soon computer manufacturers, enjoying their own explosive growth, were hooked in to frozen food sales and in two or three years the larger suppliers' depots had computerized methods of recording sales, keeping sales records, and producing sales invoices all at the local depot location. This development was worth while because of the growth in the ownership of deep freeze cabinets in the home as well as in shops, hotels, restaurants, etc.

The final act of this particular stage of the development of distribution took place when the pressure of delivering to supermarkets and similar locations, in virtually every town, led to the concept of bulk deliveries using specialized vehicles carrying load handling equipment such as tail lifts.

In an attempt to reduce the weight and bulk of plate eutectic systems, tube systems were developed and some manufacturers paid serious attention to the development of air-blown systems for vehicles, although these were not yet universally accepted for local delivery routes. The need to operate expensive refrigerated vehicles on more than one shift (the limit of the eutectic systems) was recognized. Multishift operation was experimented with, and this required 'on the road' refrigeration for the eutectic systems, or reliable engine-driven air-blown systems.

2.4   Developments in the 1980s

This decade has shown very clearly just how distribution systems are of necessity controlled by the market place. Almost all the large producers of frozen products have now moved from in-house distribution to a contracted service with a general distribution specialist. Indeed, the emphasis has swung totally from the manufacturers to the retailers; the end of the 1980s has seen the largest retailers developing their own distribution centres nationwide, giving dedicated back-up to their own shop chains.

This change was not immediate. For the first five years of the decade, some manufacturers attempted to satisfy their larger customers by stocking and distributing products manufactured by others. This was only an operation to stem the tide of change for long enough for the extent of the change to be understood. In the last five years of the decade a complete reversal occurred from manufacturer-based distribution, to distribution managed by specialists on behalf of retailers.

A huge swing of this nature, involving a total change of attitude and very heavy investment by retailers (or others on their behalf) was facilitated by and indeed needed for its success, the domination of the retail business by a small number of very large companies. At the time of writing (1988), the events described above were illustrated most readily by the largest retailing groups. Their success is dependent on the willingness of local planning authorities to allow the building not only of very large out-of-town hypermarkets but also of even larger, strategically placed distribution depots.

The distribution depots are usually sited near to significant conurbations with good access to the motorway system. Siting shops presents more problems in terms of traffic congestion, but they are normally sited just outside major towns and cities. They feature large car parks and aim to provide one-stop shopping for the more prosperous members of the population who own a car. Bus services to these hypermarkets are provided, and in some instances free buses from outlying residential areas are provided to improve the catchment area.

There has been a steady growth in the tonnage supplied through frozen food wholesalers. The wholesalers' customers range from the hotel and catering industry to the small cash-and-carry shop. A feature of this business is the rapid turnover of a limited line range. The operation is more one of transport than storage, the wholesalers' cold room being a transit shed rather than a cold store.

There has been a second resurgence of franchising, particularly in ice-cream, for which this is a very suitable form of selling as the customer can be persuaded to be brand-loyal. The first resurgence was in the mid 1960s, when the major ice-cream companies moved out of owning retail businesses, and the 'Mister Softee' and 'Walls-Whippy' franchise chains were established. In the second resurgence the 'general trade' was transferred to franchises in specific areas. The general trade is, as its name suggests, all the small and individual accounts like beach concessions at the seaside. National accounts were in the main retained by the sales department of the manufacturers, and these continued to be serviced by the specialist distribution companies.

To complicate matters, distribution companies, operating in the conventional way, on behalf of manufacturers, could also be associated through a common owner with a contract company operating the dedicated depot for the major hypermarket owners.

It has been said that 'a week is a long time in politics'. It could equally be said that at the end of the 1980s a year is a long time in distribution. How were these larger changes managed in so short a space of time? Equally important, how were the large daily tonnages handled at the new centres? The answer to the first question is that, in the main, changes were accomplished in the first instance by the regrouping of companies. This provided a core of knowledgeable management which could then grow as required. The second question requires a more detailed and lengthy answer.

The two criteria which determine the scale of a distribution store are *line range* and *stock cover*. The quantity of stock carried is implicit in the second of these, in that, if it is decided to carry two weeks' stock (a typical figure) then the quantity becomes a function of the daily throughput which in turn sweeps up other factors such as area covered, number of customers, etc.

The line range determines the configuration of the store. Every line has to be accessible and so a large line range requires more access, in the form of 'faces' from which picking can take place on a daily basis, than a small line range.

Determining the quantity of stock carried for each line is an arithmetical calculation based on either known or anticipated sales. Clearly, fast-moving stock will require more space with, perhaps, the order-picking area replenished frequently from bulk back-up stock on site. Computer programmes now exist to permit complex store operations to be controlled, with the location of every pallet predetermined on delivery to the store from the manufacturers, its movement in the store controlled, and its replacement anticipated.

In keeping with the growth of the supermarket and 'superstore' business has been the decline in the order-picking of packets for retail customers. In the last five years case-lot delivery has grown from 55% to 70% of all frozen food deliveries. (In this context, cases are the outer containers used by manufacturers in forming pallet loads.) We might conjecture that very soon only the franchisers of ice-cream and the wholesalers of frozen food with limited line range will consider breaking open cases at their depots to deliver packets. This would mean that traditional distribution and dedicated distribution as outined above would be only in pallets and cases.

Looking in general at the way in which these changes have been accommodated, it is convenient to subdivide the depot operation into intake, storage, and despatch.

The intake of goods from frozen food manufactures is now universally on 1 m × 1.2 m timber pallets. These arrive at the depot in an insulated and refrigerated trailer usually 20 to 22 pallets at a time, although with some products 24 pallets are feasible. The constraint in haulage may be weight, but is more likely to be product overhang on the pallet base, making close packing of the pallets on a legal length limit of 12.2 m difficult. The vehicles reverse onto a dock leveller, usually surrounded by a dock shelter which seals onto the rear of the vehicle using pressure pads. The product being unloaded is thus isolated from the outside conditions. The vehicle is then usually unloaded using battery-powered hand pallet trucks. Occasionally so-called 'slug' unloading is used but this remains uncommon. In slug unloading the whole load is moved at one time onto the bank or some receiving conveyor system. The product is normally checked at this point for quantity, quality, and damage in transit.

In the most common systems the powered pallet trucks take each pallet, after checking, into the store and deposit it in predetermined positions, usually at the end of the storage racks, ready for the next movement. In some stores pallet conveyors have been installed which take a pallet from the loading dock and transport it into the store through a minimum-size door (Figure 2.3); within the store the pallet conveyor has its driven rollers so arranged that it can accumulate a full vehicle load, thus making for fast unloading of an individual vehicle. Pallets are taken from the end of the conveyor and placed at the ends of the racks as before.

Figure 2.3 In-loading pallet conveyor inside a store.

Storage can now be achieved in three ways:

- By using reach trucks working in wide aisles and putting to stock on racking
- By using turret trucks working in narrow aisles to achieve the same result
- By using cranes in narrow aisles

This has been adopted by several companies in continental Europe, but as of 1988 only one low-temperature installation near Bristol was working in the UK. The choice of which system to use is based foremost on distribution factors—throughput, line range, stock level—but almost as importantly on investment decisions—capital availability, site availability, forecast of use and life (see Chapter 8).

Each individual installation requires detailed study but, in distribution terms, the following comments apply:

- *Reach trucks in wide aisles*   Well proven. Slow at the fourth stacking level and very slow at the fifth. Reliable, easy to get replacement equipment. The least economic in space terms. The cheapest capital cost.
- *Turret trucks in narrow aisles*   Fast at all levels up to five. Reliable, but replacement equipment is specific to the location. Economic in space terms, more expensive than reach trucks.
- *Cranes in narrow aisles*   More economic at levels above nine. Competitive in capital cost with other systems per pallet stored at eleven high or more. Reliable and economic in space utilization. Can be wholly automatic in operation, but require sound and dimensionally accurate pallet bases.

All three systems require pallet racking as part of the system. In the case of cranes, devices are available to reach to a second pallet if the first space is empty. Alternatively, with limited line range it may be sensible to use racking fitted with gravity rollers so that loading and unloading at opposite sides of the rack is possible and there is always a pallet available at the offtake face.

In the interests of safety, potential users of racking systems should always take the advice of a qualified structural engineer on their installation. From the pallet racking, pallets can be withdrawn for break-up to compose customer orders. Some very simple methods are used, particularly with case-lot picking. The most simple and effective is to perambulate with a pedestrian pallet truck around low-level racks, say two high. This gives picking at floor level with one replenishment to hand immediately above.

Where cases are of relatively small dimensions, and particularly if they are to be opened for packet assembly into customer order, it becomes feasible to use low-level racking and a gravity roller conveyor; these can be five high and still reachable. The length of each rack is limited only by the range of lines or size of the building. Goods are picked from the faces of the racks onto powered conveyors which conduct the goods outside the store for the order to be checked.

On the loading bank the load for the customer has to be reassembled for transport. This reassembly may be into paper sacks, wheeled cages, or remade pallets. The distribution vehicles standing at the bank dock levellers are loaded either by pallet truck for pallets, by hand for wheeled cages, by gravity conveyor into the vehicle for paper sacks; cases have to be stacked by hand.

In recent years loading banks have become enclosed and well lit; they may now be the only workplace where much manpower is used. Another factor in the improvement of loading bank facilities has been the extension of working hours in distribution. Multishift operation is quite normal. A typical pattern would be: intake early in the day, then stock consolidation, an afternoon picking shift, perhaps for vehicles being exchanged at outbases, followed by the main picking shift during the night during which the bulk of the fleet is loaded. Deliveries are still usually made during the day, but the pattern is changing at some of the larger customers where intake at the shop has become continuous for up to twelve hours.

A vital development in the mid to late 1980s has been the integration through computers of sales and distribution. Some companies are now fully computerized from order capture to stock replacement. Integrated in the computer program is a traffic planning exercise and a vehicle loading guide. Linked to stock replacement is a cold store stock and product location program. The final phase—communication direct between the computer of the retail customer and the distribution company—has led to a reduction in sales personnel engaged in the industry. The tendency to fewer but larger customers has also had an effect here.

2.5 Present requirements

The discussion in previous sections has illustrated the way in which the industry has moved from its beginnings in manufacturer-controlled distribution to retailer-controlled distribution. What does this currently mean?

At the end of the 1980s the distribution pattern falls into three recognizable groups:

- Major retailer with own control distribution centres
- Distribution by specialist company to individual centres
- Distribution by franchiser or wholesaler to smaller outlets

The first of these requires only the delivery to the distribution centre, in an agreed pattern, of whole pallets of products. The central distribution site may be operated by the retail chain or by agents on their behalf. Breakup of bulk stock and the preparation of shop deliveries is carried out at the centre. Intake to the centre and despatch of the assembled shop order is in the hands of the same management team, and economic practices can be made to prevail: this means controls such as minimum order size and minimum line order. In the case of pallet deliveries, a control could be a minimum number of pallets. Very soon, for a large retailer, the criterion becomes number of drops per vehicle. Where the distribution centre has multitemperature storage and ambient storage, the criterion can become one of filling a multitemperature vehicle per shop. Carrying frozen and chilled products in one vehicle is totally feasible, and is practised in several large operations. Carrying frozen and ambient product on the same vehicle has been done, but is less common. Vehicle design will be discussed later in this chapter.

Major retailers are unlikely to be interested in handling anything smaller than pallet loads reassembled from bulk after a case picking procedure, or wheeled pallets containing picked loads of cases. It follows therefore that their distribution store will be designed for rapid intake of bulk pallets, enough storage for defined stocking period, and the facility to despatch pallets or reassembled pallets easily.

Control of product temperature has led to the use of temperature-controlled loading banks and enclosed ports at which bulk supply vehicles are emptied and distribution vehicles are filled. The product passes through the loading bank area as quickly as possible. On the intake to the cold store this can be achieved by the use of powered roller pallet intake conveyors. On the despatch side, order-picking is carried out in the store, final order assembly and checking against the invoice is carried out on the loading bank, and the vehicle is loaded. At a large site parts of one order to one shop location may be picked in separate parts of the depot and leave the stores by separate exits to the loading bank. In these circumstances the whole order can only be brought together on the bank and checked before loading.

Product may progress to the loading bank through several routes, depending on throughput of the line. Very fast movers may be moved in whole pallets—a relatively simple process of retrieving a pallet from the bulk stock by reach or turret truck. Lines with the next category of priority can be readily picked from ground-floor pallet lines by a man perambulating a pallet on a hand-operated battery pallet truck. Slow-moving lines can most readily be picked by multilevel order-picking trucks. In these trucks the man travels with the forks, as high as five levels, selecting cases. The necessarily slow travel times make this method rather slow but it has its place in the scheme of things, dealing with the lines which must be stacked but which are not sold in large quantities in any individual order. Thus the final order on the loading bank may arrive through three doors. It should all pass one point for checking before loading into the vehicle. At many large distribution centres the loading banks, for the reasons outlined above, are the busiest places on the site.

In the second case of distribution by a specialist company to a wide variety of individual sales outlets, a further method is required to cope with loads to the smaller outlets where whole cases cannot be sent. At these centres all those activities described under the previous section as appropriate to major retailers are carried out, plus a 'broken case' section. This section takes

Figure 2.4   A 'broken case' unit with two gravity feed conveyors either side of a double rubber belt conveyor.

considerable space, often for only 30% of the total throughput of the cold store. One method of carrying out this function is to present whole cases in gravity feed conveyors in frames perhaps forty units long and five high (Figure 2.4). Two such units either side of a rubber belt conveyor would allow a man on each circuit to pick 400 lines. By having two belt conveyors down the middle of the walkways between the frames, one on top of the other, two men can pick two orders simultaneously as they follow each other along the frame face. The conveyors discharge in the loading bay for checking and accumulation into despatch loads. If very large ranges of lines have to be picked, a system of this kind may even have to be used for whole-case picking.

Because there is not a great deal of pallet movement required to a picking frame of this type, once it is loaded (usually on a previous shift), the frame can be positioned on a mezzanine floor inside the cold store permitting case-picking by perambulating pallet trucks underneath. Care must be taken with such an arrangement not to obstruct air flow around the store. Since most cold stores have their air coolers at high level it should be noted that a mezzanine floor at the stores' half-height may be cooler than the store average temperature because it is closer to the coolers.

It is worth mentioning at this point that a retailer operating his own distribution can control his order-picking by the simple, if crude, means of restricting the line range he carries. The specialist company is normally expected to carry all line range of the manufacturer for whom he provides a service. It is a simple point, but the reality is that manufacturers do tend to have a long 'tail' to their product range. The 80/20 rule, as it is called, seems to be general in food manufacture: 20% of the line range comprises 80% of the volume handled.

The final situation, distribution by franchiser or wholesaler, reverts to the small scale of the first situation, in that the management are at liberty to either extend or curtail the line range carried according to the space available for stock. The methods chosen are in consequence usually quite simple, often involving no more than manual picking from a 'face' of pallets onto a single conveyor feeding out to the loading bank.

In almost all cold stores used in distribution which are also intended to have stock capability, a universal feature is the placing of pallets in racking. This serves several functions. It minimizes the use of volume by using height, it makes pallets readily accessible, and, of considerable importance, it makes stock rotation feasible. The number of racking required is an arithmetical function of the number of lines, the throughput per line, and the stocking period. The second universal feature is the use of part of the store for marshalling or order-picking. Depending on the type of store these phrases mean a range of activity, from the picking of a single shop order to the assembly of picked pallets into a vehicle load.

All this activity requires very good lighting by industrial storage standards, because labels have to be read and invoices checked against consignment notes, etc. In the store 250 lux is normal in the aisles of the racking and order-picking areas. On the loading bank 350 lux can be expected.

Two-way communication is important when picking and checking teams are working at the same time inside and outside the store. Items may be missed, and the checker must be able to call up the picker and get the order completed. Some companies have also installed two-way radio in the cabs of high-lift trucks, where these are fitted. Cabs on trucks are feasible, but by no means in common use as the heating system is such a drain on the battery. Only those companies with high throughputs, who have regular battery changes in consequence, and who also probably have a permanent spare machine, can really justify the added complication of heated cabs. This does not prevent a conventional truck being fitted with a small loud hailer and a radio.

Implicit in the discussion so far has been the use of pallets as the sole means of transporting bulk into the store. This situation is true of a large variety of distribution stores visited by the author throughout Europe. There is some variation in actual pallet base size, with the 1.0 m × 1.2 m pallet being most common. Once bulk is broken down, however, although the frozen industry prefers cardboard cases operating over roller conveyors, some products are packed in other containers, such as paper sacks and varieties of plastic bags. The system chosen must accept reasonable variations of these.

Other industries carrying out similar functions, but in ambient conditions, have developed other containers—for order-picking in particular. The use of bins which recirculate on conveyors between picking faces and final packing is well established. They are very useful when product damage is prevalent, as the bin system can be designed to take the wear and tear of the make-up period when goods are being accumulated from sources all around the depot.

When automatic bulk handling is required some other industries have chosen different options such as once-through pallets, plastic pallet-type bases, and, to cope with damaged pallets, slip boards on which the pallet is placed before entering the system. Even when pallet-pool systems are in use, pallet base damage can be an embarrassment with automatic systems. At two automatic high-rise cold stores in Europe, taking in public goods, it was estimated in 1987 that almost 30% of the intake either had to have a further sound pallet base placed under the existing one, or the pallet had to be repacked. One of the stores even had a permanent hydraulically operated machine in the loading bank to facilitate this.

More detailed consideration can now be given to vehicle loading. There is a marked difference between handling of chilled goods and frozen goods. Chilled goods are mechanically less robust and therefore much more readily damaged. Meat pies and soft wrapped sausages illustrate the point. To ensure that these products arrive at the customers' premises undamaged it is normal practice to protect them by packing at the point of manufacture in plastic trays which are carried in shelved cage pallets. Only at a point near to the final distribution, usually, for example, the shop delivery van, are these trays reloaded into shelving in the van. Frozen goods are robust, and packeted goods, perhaps with only an overwrap of polythene, can be packed in a paper sack and stacked up to 1.8 m high in an open van without damage—provided of course, that the temperature maintained is −18°C or less!

This method, which can also include re-use of surplus cartons from the cold store order-pick, is satisfactory for small orders. When larger quantities are involved, it becomes necessary to use either wheeled cage pallets or wooden pallets. Wheeled cage pallets are expensive to purchase and maintain and require a lot of space when idle, even if they are of the stacking variety. It is easy to see why they are unpopular with distributors on these grounds, but popular with some supermarket chains because of the flexibility they give at the shop. Wheeled cages can be readily packed with cartons, sack products, and packets. A simple stretch film can be applied by hand to prevent spillage or pilferage of goods. At the shop the cage can be wheeled to the display unit and unloaded into it without any other handling or equipment being required. Wooden pallets can be packed by hand on the loading bank—some skill is required, but this is soon acquired with product familiarity. When the pallet has been packed up to 1.8 m high it can be stretch-wrapped by hand, or preferably by machine, to give it stability and security. Cardboard corners are sometimes incorporated.

Some distributors also use shrink-wrap methods. In this technique a plastic sleeve is placed over repacked pallet and shrunk on by a hot air stream. The process takes a few seconds and there appears to be no heat shock to the product. These plastic sleeves have a top and so the pallet is totally secure from pilferage unless a knife is used to cut the plastic; this is then very obvious. In the vehicles themselves it is unusual to see shelving used except for easily damaged chilled products.

2.6 Distribution vehicles

By common consent vehicles now offer full utilization of floor space, and in northern European countries it is almost universal for the vehicles to be designed for the driver and loader to enter the refrigerated space. Some ice cream companies selling elaborately made ice-cream combinations use side-hatch vehicles on account of the close temperature control that is possible with this design, but they are now a minority.

Distribution cost pressures, which have led to larger vehicles, bigger drops, and faster turn-round, have now also forced multi-shift operation and this has had the most profound effect on design. For many years the distribution of frozen goods was a single-shift daytime operation, because only at this time did the customer have staff present to receive the goods and check them in. In consequence, maximizing vehicle use and mileage was not the major objective in vehicle design. Reliability was of prime importance, in the sense that vehicles should be available every day scheduled at the correct temperature.

Since the vehicle was on the road all day but available all night at the depot, it was quickly seen that the most reliable system of refrigeration was that provided by the eutectic plate. By the start of the decade this method was virtually universal in shop delivery vehicles not only in the UK but throughout Europe. Eutectic plates were carried fixed to the walls, and later the roof, of the vehicle. Recent versions of this system consisted of steel tubes and steel beams trapezoidal in shape (Figure 2.5) and finally plastic beams. The development of roof-mounted system was, of course, intended to keep the floor area clear for pallets and wheeled pallets.

The disadvantages of these systems were the weight of the eutectic filled parts plus the weight of the condensing set carried on the vehicles to charge the system during the night. A ten-hour working day required ten continuous charge hours during the night. As commercial pressures forced the multishift use of vehicles these systems were made viable by on-the-road equipment such as generators driven from the vehicle gearbox power take-off, which provided power for the condensing set. Hydraulic drives were similarly available.

The operators were reluctant to abandon the reliability of eutectic systems. However, the industry has developed along the path of larger drops and multishift working, and this has inevitably led to larger vehicles with heavy payloads. To meet this market the transport refrigeration suppliers developed a more powerful yet compact plant, and a change to air-blown system was made. This equipment has the disadvantage that it is noisier than the older

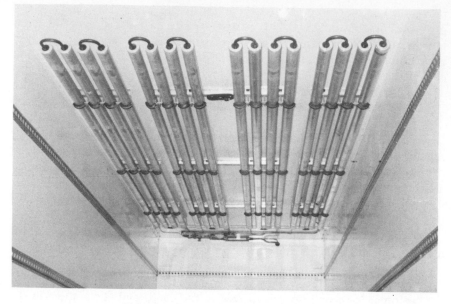

Figure 2.5   Eutectic plates with steel tubes.

methods. In a noise-conscious society some distributors have even produced systems combining air-blown and reduced eutectic systems. Using these, the diesel drive for the compressor is in use on the road and the vehicle lives off perhaps four hours eutectic standby when waiting in noise-sensitive areas or when loading and unloading. An independent diesel-driven compressor gives total flexibility of operation, but distribution depots are not immune from the requirements of planning authorities and 60 to 70 vehicles loading at night can be a problem in this context.

Reference has already been made to the switch of influence over distribution from the manufacturer to the retailer. Two further manifestations of this are the requirement to fix delivery times at large locations, and product off-loading. In the case of fixed delivery times the retailer allocates a slot of time during which the delivery will be accepted. If this cannot be met by the supplier the vehicle has to be re-presented at another arranged time. This can work well when it is properly planned, but takes little account of the seasonal vagaries of travel in busy areas like northern Europe.

The case of product unloading is more interesting. It is very unusual for the retailer to provide his own vehicle unloading facilities. In consequence, larger vehicles delivering to the larger account carry a tail lift to permit the unloading of pallets and wheeled cages (Figure 2.6). These tail lifts are usually full width,

Figure 2.6   Tail lift delivery vehicle.

Figure 2.7   Side-hatch delivery vehicle.

mounted at the rear of the vehicle. To permit reversing up to the loading bank a version of the tail lift has been developed which tucks away under the rear of the vehicle, thus permitting it to travel short distances in reverse with the lift clear of the rear doors and with sufficient ground clearance, usually 150 mm. Other tail lifts, half or one-third vehicle width, have also been developed for off-loading in the street, for the larger shops which can take wheeled cages but still have access only across the pavement.

A feature of all tail lift designs must be that they are fail safe. In other words, a hydraulic leak or failure of lifting chain or wire does not allow the lift to drop to ground level from the 'park' position.

Over the years various methods of unloading vehicles sideways on to the pavement have been tried. These are, of course, mainly of interest when the shop can only be serviced over the pavement, still a frequent situation. For small deliveries these methods include side-hatch vehicles (Figure 2.7). Vehicles with driver access via a front vestibule have also been used (Figure 2.8). These require the vehicle to be loaded on a first off/first loaded pattern, the reverse of the more usual situation.

Finally, large vehicles can now be fitted with a rear-mounted tail lift which slides away as a complete unit under the floor of the body. All rear-mounted devices of this type have to include in their design the rear bumper legally required in the UK and defined in the Ministry of Transport's Construction and Use Regulations.

Figure 2.8 Delivery vehicle with a front vestibule for driver access.

2.7 Present-day materials for construction and equipment

The evolution of the marketplace, referred to on more than one occasion in this chapter, has led to evolution in the concept of the distribution centre. A large retailer now builds a centre to which all his suppliers deliver. Its geographical location is determined by transport studies. The choice of site may of course be limited by planning controls, but fundamentally it will have access to the motorway system and it will serve a slice of the population numbering between 5 and 7 million. Since all suppliers deliver there, it must operate at the four defined temperatures common at large retail outlets; frozen at $-25°C$, cold chill at $0°C$, chill at $+5°C$, and ambient. As a generalization it will have two weeks' stock of products (but not of course of some perishables!).

The challenge is to build such a complex building in such a way that it is effective when commissioned and also adaptable if there is a significant change in requirement. One answer has been to accept the unchanging elements and build these in a permanent configuration, and allow for those elements subject to market forces to be flexible. For example, the yard, lorry parking, car parking, vehicle circulation roads, and vehicle maintenance can be reasonably defined and are a function of the tonnage carried rather than the nature of the tonnage. It is necessary to take account of existing legislation, and of any legislation known to be in the pipeline. The influence of EEC directives is very important as the harmonization of Europe in 1992 comes closer to reality.

In a similar way, office and amenity accommodation can be planned on the basis of throughput and hence numbers employed. Accommodation for mobile materials handling plant requires careful thought, but, in principle, provision of repair facilities, battery (or engine) maintenance and charging that is suitable for today's systems is likely to prove adequate for tomorrow's system's.

The future variations are likely to be in size relationship of the temperature-controlled rooms and their throughput, and the changes necessary in the loading-in and loading-out facilities to meet changes in throughputs.

To summarize, if the various temperature-controlled rooms are created within a single large internal envelope, formed in a large structurally independent shed, it becomes possible to accommodate change without structural modification. This is made all the more possible today by the existence of machines capable of manufacturing continuous steel-faced insulation panels. Virtually the only limit to the panel length is the ability to transport it by road. It may be necessary to divide a very large steel-framed building with one or more fire walls, but usually these can be placed where they do not obstruct the economic design and layout of the cold rooms.

Within the shell and fire walls, chambers can be constructed using factory-made panels, some of which can be demounted and moved if necessary. Only the floor design requires special consideration, in that any chambers below freezing point require insulation and a frost heave prevention mat (unless the floor is a suspended one). Clearly it is cheaper to provide any areas which may be used for frozen storage with heater mats and suitable insulation during the initial build rather than later. Some recent designs have covered this point by providing 50% of the floor area of the chill facilities already prepared for future frozen use if required.

Temperature-controlled rooms in distribution stores are still designed as rectangular boxes with the height governed by speed of working from the floor by mobile reach or turret trucks. This is to retain flexibility. Only for longer-term storage of factory pallets does the industry seem to countenance purpose-built crane-operated stores.

Access to and from temperature-controlled stores is usually by means of air-operated sliding doors. Air operation, although requiring the local provision of compressed air, is seen by operators today as giving shorter door opening and closing times and greater reliability than electric drive systems.

In-loading docks may also be used for out-loading. They should be as spacious as possible because of the high activity levels. They are working areas, and should be as hygienically finished as the stores. They should also be well lit and temperature controlled. As well as acting as a buffer between temperature-controlled areas and ambient, the loading bank is often the place where delivery and reception problems are cleared up. A compromise temperature to cover these requirements is $+5°C$; goods being handled through a zone at this temperature will be compatible with cold chill and chill rooms, and the advantage for frozen rooms is that it stabilizes the ingress heat load of door openings. Note that coolers on the loading bank providing this temperature should have defrost facilities, as the coil temperature may have been below $0°C$ to achieve a general $+5°C$ in such onerous conditions.

The comments above apply equally well to out-loading. The significant differences are that the goods being out-loaded have been under control of the distribution centre management and so should be at the correct temperature and undamaged. The packed form may be cardboard boxes, wheeled cages, wooden pallets, or other special types of cage. Either way the loading dock must also be able to out-load a variety of vehicles from 10 to 38 tonnes. This is not difficult, as dock levellers can cope with the differing floor heights. As a guide the dock leveller should be long enough to keep the slope down to the lowest vehicle no more than 10%. Dock levellers can be electrohydraulically operated or manual. They can have a fixed lip, a folding lip, or a hydraulically operated lip. The more vehicle movements there are, the better is the case for hydraulic assistance to raise and lower the leveller.

In recent years the sealing of the vehicle dock to the vehicle has improved. Plain rectangular-section seals are still used, of course, particularly if the dimensions of the vehicle using them are known and unlikely to vary. Where a wide variety of vehicle door openings have to be matched, seals which slide on runners are available, as are seals with an irregular section which fold to fit narrower or wider vehicles. Most recently the development of air bag seals has made universal sealing of all sizes of vehicle aperture a reality. These differ in

Figure 2.9 Dock shelter with an up-and-over door.

operating principle in that they do not rely on any form of compression between the vehicle and the dock shelter. The vehicle is positioned and the air bags are inflated round the aperture and seal on the outside surface of the vehicle. Although very effective they are the most expensive of the four systems and require a compressed air supply.

Finally, it is usual for the whole dock shelter to be sealed off from the loading bank by a door. The most favoured design is a segmented up-and-over door. This can be counterbalanced, and manually or electrically operated (Figure 2.9).

The position of the dock leveller requires some thought. The majority of designs in the past were simply a development of the old open loading banks. Dock levellers were usually hung from the leading edge of the dock, with the port built round them. With yard space at a premium it is now more usual to make the front of the building the dock pad line and have the levellers indented into the bank. This does raise some safety questions regarding the overturning of trucks on depressed levellers. This can be overcome in part by having guide rails to prevent trucks crossing levellers transversely, and in part by having fixed park positions for levellers. This interface between the loading bank and the vehicle is absolutely crucial to good operation, and time and money is well spent on getting it right for the particular operation.

The type of enclosed dock port described earlier raises a number of ancillary issues which affect the operation. The loading supervisor cannot see the vehicle, so a system of signals, usually minature traffic lights, should be provided to tell the driver of the vehicle that a port is available. He may have a tail lift which is not retractable. The only sensible way to cope with this is to provide an over-wide space under the dock leveller so that the driver can lower his plate and reverse onto the pads with the tail lift plate parking under the dock leveller plate. To achieve this, and also as a sensible maintenance provision, the electrohydraulic unit of the dock leveller should be in a position where it can be reached and serviced without anyone crawling under anything!

The creation of a protected environment for the product during loading has totally enclosed the vehicle's rear end, and it is necessary to provide lighting which will illuminate the vehicle interior without getting in the way. Small spot-lamps with a defined beam are usually a reasonable compromise solution to this problem. In the event that the vehicle has to remain at the bank for an extended period, an electrical connection should be provided at the side of the port door to permit operation of the refrigeration plant without running the diesel engine. Usually a 15HP plug unit is provided.

2.8   Limitations of choice

Distribution stores and vehicles are usually the centre link in the chain from manufacturers to retail outlet. In consequence the distributor is forever trying to look forward, and over his shoulder at the same time.

He must look forward to provide the shop services required, and this may incorporate multitemperature delivery in a variety of delivering modes, anything from whole pallets for fast-moving products to individual parcels of expensive products like frozen salmon or liqueur ice-cream. He must look over his shoulder so that he is not surprised at changes in the form in which his bulk goods arrive. Catering lines tend to gravitate into larger unit size as soon as a particular line increases in popularity.

The distribution contractor must clearly retain flexibility and this often means keeping to basic shapes simply so that they can be readily changed. Imagine trying to connect a purpose-designed pallet store built as a tower with cranes into order-picking of cardboard boxes stored on plywood slips. This is why distribution stores are invariably rectangular, often with a length about twice the width and no more than 11 m high (Figure 2.10). This is a practical limit for the use of reach and turret trucks. At this height, reach trucks need side-shift mechanisms on the forks and even turret trucks operate more slowly due to the effects of mast sway.

The distribution contractor has to retain adaptability for his investment. Fortunately, although building design is constrained to be simple on this account, in the area of storage temperature the ranges acceptable are much more clearly defined. This means that plant can be designed specifically for a known product load in a defined temperature band.

Agreement can be obtained, from supplier through to end-user, on the maintenance of a cold-chain procedure in which the distributor plays his part

Figure 2.10   Distribution store to show typical dimensions suitable for pallet storage and truck access (see text).

and an important segment of the disposal cycle is controllable. By negotiation, delivery can often be throughout 24 hours which makes it simpler to avoid traffic problems and represents a reduction of limitations, but the choice of any specific time for delivery, no matter how convenient, does impose a limitation on the effective use of a vehicle and tends to negate most of the potential benefit.

The limitations outlined above are negotiable, but some limitations in vehicle size and shape are fixed by statute. The distributor has to resolve the load he wishes to carry into an equation whose variables are width, length, axle weight, and number of axles. It may be that his decision is not limited solely by these considerations. His customer may place further restrictions on him for reasons of access to the site, or in pursuit of a good-neighbour policy. Noise is of increasing importance each year and planning approvals for new developments frequently define maximum noise levels at the site boundary.

One of distribution's long-standing problems with smaller vehicles, namely the overloading of the front axle with unloading from the rear of a two-axle vehicle, was eased in 1987 with the adoption of a 17-tonne gross vehicle weight for two axles, the rear drive axle having an approved maximum capacity of 10.5 tonnes. This size of vehicle is very popular in distribution, as it is widely thought to be the maximum acceptable size for street delivery in towns.

From 1989 the maximum insulated box used for delivery in the EEC will be 12.2 m platform length by 2.52 m wide. This latter dimension is defined from a maximum legal exterior width of 2.6 m where the sidewalls are not less then 40 mm thick. A vehicle of this size can carry 32.52 tonnes on four axles or 38 tonnes on five axles. These are gross vehicle weights, not payload. A vehicle operating at these weights and sizes requires specific unloading facilities and could not be considered as suitable for universal street unloading.

The type and protection afforded by the packing used can also be considered as a limitation on the distribution operation. Skimpy packaging has to be more carefully handled, and liability for damaged goods becomes a more contentious item in the distribution deal.

2.9   Regulations and standards

In the distribution industry, there is a substantial difference between regulations and standards applicable to cold storage and those applicable to refrigerated transport.

The structural element of cold storage has of course to meet the same Building Regulations as any kind of building in the UK. The cold storage envelope and refrigeration has only to meet the technical requirements imposed by the end user. To illustrate this point, a cold storage operator who wishes to store products under the EEC intervention regulations has to have his store certified by Lloyds Register because this is required by EEC regulations. An operator who is only handling his own products can set his own standards. Having said this, there are a number of standards and codes of practice that he would be ill-advised to ignore.

If we assume that in designing the store the operator has satisfied the local authority with the structure and building, and the fire officer with the fire precautions, may still have to meet further requirements detailed by his insurers. These may include, for example, a sprinkler system inside those areas of the structure above 0°C and most likely a sprinkler system enveloping the low-temperature areas externally.

The next requirement is to design and install the insulation envelope, and here an operator would be well advised to use the code of practice published in 1986 by the Institute of Refrigeration entitled *Code of Practice for the Design and Construction of Cold Store Envelopes Incorporating Prefabricated Insulating Panels*. Prior to the introduction of this code it was necessary for the constructor to agree all details with his client, and it was usual for the major cold storage companies to have their own in-house standards for the various elements.

What the 1986 Code of Practice has done is draw together good practice from a wide cross-section of the industry and provide a sound basis on which an individual design can then be tailored to suit specific needs. The code should be regarded as the essential starting point for any design and its use should always be stipulated in any enquiry; it does not preclude specialized requirements being added to the design.

A further safeguard frequently used is to stipulate that the store will meet the requirements of Lloyds Register. This is an internationally recognized certification procedure controlled by the Refrigeration Department of Lloyds Register of Shipping, and in essence is a test of fitness for use.

The refrigerating machinery and its design can be required to conform to a wide cross-section of standards covering materials, manufacturing methods, welding standards, and similar matters. Of particular interest is the fact that, in the UK, safety in refrigeration systems has its own standard, BS 4434, and no plant in the UK should now be constructed outside its recommendations. The British Standards Institute works in collaboration with the International Standards Organization whose role is to promote and prepare standards acceptable to all members and countries. This organization bridges political barriers, and eastern European countries have a positive role in the committee structures reviewing existing or intended international standards.

Most western industrial nations have standards, in some part at least, in the wide area of design and installation now to be found as part of the refrigeration engineer's working life. Outside the UK perhaps the most notable are the American, German and French standards. Indeed, in the 1960s, because the UK had not progressed very far at that time with standards in refrigeration, it was quite common for a plant in this country to be built to American standards. This was also, in part, a recognition of the size of the American refrigeration industry's home market.

In recent years much hard work has resulted in progress in the field of standards. The Institute of Refrigeration has produced Codes of Practice on the *Design and Construction of Systems using Ammonia as the Refrigerant* and on commissioning inspection and maintenance of such plants. The Institute has also produced a code for the design and construction of refrigerating systems using chlorofluorocarbons (CFCs). These have all appeared between 1979 and 1984. The existence of standards and codes of practice makes their use implicit where appropriate in order to satisfy the general approach of the Health and Safety at Work Act. It could be said that there is no legal requirement to use specific standards in design, but the general duty of care in the Health and Safety at Work Act makes the position of anyone choosing to ignore these works legally precarious to say the least!

In the UK refrigerated vehicles have to comply with every detail of legally enforceable regulations generally known as the *Motor Vehicles (Construction and Use) Regulations 1978*. Other regulations are enforceable in special instances, or for special vehicles. In addition, in every fleet of vehicles within a Licensing Authority's area, a person must be designated to be in charge of the vehicles and that person must have a Certificate of Professional Competency. Implicit in the gaining of such a certificate, by examination, is the knowledge of the legally enforceable regulations.

The regulations are so comprehensive that any form of summary would mislead. It is sufficient to say that all aspects of size, weight, vehicle loading, axle configuration, safe braking, noise, drivers' hours, etc., are regulated and the best introduction is probably to read a publication such as the *Freight Transport Association Handbook*. One item of interest is the easing of the control of overall width in 1989 from 2.58 m to 2.6 m, for insulated vehicles where the side wall thickness is not less than 40 mm. The effect of this is to permit the carriage of two international pallets of side 1.2 m with their long sides across the vehicle.

Refrigerated transport travelling across national boundaries must comply with the conditions of the Agreement on the International Carriage of Perishable Foodstuffs (known as the ATP Agreement). In the UK the provisions of this agreement are enforceable under the provisions of the *International Carriage of Perishable Foodstuffs Act 1979*. Numerous other regulations affecting international vehicle operations are enforceable, and comprehensive information on these is available from the Freight Transport Association.

2.10 Current development trends

Distribution cold storage

The most significant development in recent years has been the introduction of wholly computerized methods of handling what used to be called 'the paper work'. From the initiation of an item of business by the placement of an order, all the downstream actions can be handled by computer. These include order take, transmission to distribution centre and raising of invoices, stock control at distribution centre and stock replenishment, stock rotation within the cold store including the systematic loading of the store, traffic planning based on the goods to be moved and the location of the customer—in other words all activities having quantities and routes as constituents. For large companies with numerous outlets it is feasible first to process the order information centrally to give sales information before transferring the information for distribution purposes.

The benefit of this kind of operation to the distributor is immense in terms of speed of response to sales and control of stock. The largest retailers can have a 24-hour cycle of order to delivery, and stores can operate with hundreds of lines while carrying less than two weeks' stock. A final benefit is that the store is better utilized, and space occupancy of over 90% is claimed by some companies.

The first software packages for this type of system became available around 1980. Many firms now offer part or whole packages.

The second trend has been a movement away from regarding the distribution centre as a warehouse, to seeing it as an integrated workplace and the link between manufacturer and customer. A consequence of this change in attitude has been an improvement in working conditions for employees. Modern distribution centres have canteens, food-standard changing rooms and washrooms, loading banks which are enclosed and frequently temperature controlled, good lighting to allow parcel codes to be read or identified by code-reading machines, and walls and ceilings finished in light and easily cleaned materials. A distribution centre will have auxiliary services such as vehicle repair workshops, a mechanical vehicle wash, mobile handling plant workshops, and adequate parking for all vehicles using the site, including private cars. These sites usually operate a 24-hour cycle and so must be self-sufficient in operational needs.

A final thought on this theme is that refrigerated foodstuffs are relatively expensive. The modern systematic approach helps to minimize product loss in repacking and handling, a cost which in most cases has to be borne by the distributor.

Turning from systems and environment to hardware, development has been less spectacular. This is understandable in that the role of the distributor is usually to accept bulk deliveries from factories and remake the product into assemblies for delivery to retail outlets. In a large centre handling several thousand product lines the capital investment required to pick orders or even make up case-lot assemblies is prodigious. This implies a return on investment of several years' duration, and in this period change in the market place may render invalid the range of decisions which led to the choice of the original mechanical selection method. It is therefore understandable that the most sophisticated installations have been installed by those organizations which own and operate their own distribution systems for goods of their own manufacture: for example, some sectors of the tobacco industry, or in a general sense some mail order companies.

In the refrigerated distribution industry there is a decline in the order-picking of units as sold over the counter and a corresponding growth in the selection of cases holding a multiple number of shop units. Cases can be picked in a variety of ways, but most involve some human muscle power. Still in common use is the perambulation method where an order is made up by a single progressive walk and collection, past racked pallet 'faces'. This can be achieved at several levels using a multilevel order picker which lifts the operator to pallet 'faces' vertically as well as longitudinally. Cases can be loaded onto multiple sets of roller conveyors with an operator picking off one face and loading a take-away conveyor by hand.

Conveyor systems of great complexity and length can be installed along rows of pallet faces in racking. The operator walks each level area loading the conveyor which can be arranged to feed each order to a collating conveyor ready for repacking in a receptacle which enters the distribution vehicle. One method, the SI system, can pick cases sequentially from sliding racks onto a belt using a computer-controlled release mechanism at the low end of each inclined rack. Not all products are suitably packed for handling in this way.

Flexibility is still of paramount importance, and this requires a substantial element of manual operation. To sum up, large distributors still build cold stores and chill stores as regular-shaped cubic structures, usually fully racked for pallet holding with defined areas, either continuous or adjacent, where marshalling of orders is carried out under manual control but using mobile FLT equipment and conveyor to take the carrying element of the operation.

Progress has been made in specific areas, of course. It is perfectly feasible to use computerized robots to pack pallets with cases. There are also commercially available palletizers that will pack a pallet with predetermined case sizes. The most recent development is in television cameras which can be made to 'read' cases being fed to the robot or palletizer; this usually requires the case to carry an identifiable bar code.

Distribution vehicles

The most obvious trend in the development of refrigerated distribution vehicles has been the steady increase in size. This is commensurate with the steady growth in numbers of large stores and a drop in the number of shops taking over-the-pavement deliveries. Traditional over-the-pavement deliveries involved nothing more sophisticated than a sack barrow on which the driver manually stacked his delivery order. A suitable vehicle was a two-axle lorry carrying a payload of four tonnes. This represented a full day's work for one man handling everything himself and driving a typical route of 60–80 km.

When larger stores wanted deliveries on pallets or in cages, vehicles on two axles carrying seven or eight tonnes payload were quickly developed. These vehicles carried a tail lift to facilitate unloading. In recent years this method of operation has been carried to its logical conclusion with vehicles carrying a payload of twelve tonnes on three axles and fitted with a folding hydraulic tail lift. All these three types of vehicle operate at one temperature.

Very large superstores generate their own in-loading traffic problems. One way of alleviating them is to have maximum length vehicles, with up to three temperatures in compartments. This is not technically easy without carrying complex plant, as the range of temperatures could be $-25°C$, $0°C$, and $+10°C$. The plant must therefore be able to heat as well as cool if flexible loading is to be achieved. With the addition of heating batteries the required range can be produced:

- By liquid nitrogen injection
- By having an evaporator and heater in each of the three compartments
- By using a common refrigeration plant and ducting for cooling, with auxiliary heating batteries when higher temperatures are required.

Vehicles of all these descriptions are now beginning to appear on the roads. It is normal for such vehicles to have either a folding or a sliding tail lift, both giving unobstructed access to the rear of the vehicle when at the loading bank.

The relaxation of overall vehicle width of 2.6 m for vehicles with side walls not less than 40 mm thick has meant that long-distance bulk carriers of pallets can be loaded more easily. The effect of this relaxation on distribution vehicles has not yet become evident.

Some companies have developed specialized vehicles for their own operations, and two-deck refrigerated vehicles, carrying cages, are already on the road. One company has a development of a multilevel, multitemperature vehicle in use. This has an integral tail lift to give access at all levels.

Almost all vehicles, except those using liquid nitrogen, now have some form of air-blown refrigeration. This became necessary when multishift operation of vehicles became commonplace. The reliability of eutectic storage systems, which have served the industry well, was offset by the need to charge the system on a 'one hour's charge for each hour used' basis. Several effective on-the-road charging systems were developed, but they do introduce another level of complexity to the vehicle.

Finally, temperature control and recording is now becoming much more sophisticated with many customers requiring proof of transit temperatures between distribution centre and shop. Systems are available which monitor the plant operation, control it, and give a record either printed on the vehicle or later at the centre when the recorder's memory is transferred to a printer. There is no doubt that in time there will be EEC recommendations on temperature levels and records on refrigerated vehicles. Fortunately technical development has already progressed to the point where these requirements could be met, and there will undoubtedly be further developments in this area.

3 Controlled atmosphere storage

D. BISHOP

3.1 Introduction

Controlled atmosphere (CA) storage is used to extend the storage life of seasonal perishable products. This technique can be used for many fruits and vegetables, but it is primarily used for apples and pears.

Perishing is essentially over-ripening, and ripening is a consequence of cell metabolism. In normal metabolism in the presence of oxygen, the products of respiration are carbon dioxide, water and heat (Figure 3.1); low levels of oxygen inhibit the rate of respiration. The plant hormone ethylene is also produced during the ripening process and acts to promote ripening further. Extension of the storage life of fresh plant produce is dependent on applying techniques which reduce the rate of cell metabolism without causing injury to the produce itself. Refrigeration is the most important technique but, because produce can be frozen or suffer from low-temperature disorders, a compromise minimum safe temperature has to be accepted. Measures to delay the synthesis of ethylene or inhibit its action may also result in an extension of storage life for some types of fruit.

The most important supplement to refrigeration for the storage of fruit is CA storage, in which the carbon dioxide produced by respiration is allowed to increase to a safe maximum level and the oxygen is decreased to the lowest safe concentration [1]. Generally, fruit should be harvested for long-term storage when slightly under-ripe and in advance of the onset of the stage of rapid ethylene production. It should be cooled rapidly and the CA conditions established as soon as possible in order to obtain the most beneficial effects.

The precise levels of temperature, oxygen, and carbon dioxide required to maximize storage life and to minimize storage disorders are extremely variable. They depend on produce, cultivars, growing conditions, and maturity. Optimum storage conditions can vary even from farm to farm and from season to season. Recommendations are regularly published by the various national research bodies and extension advisers, and these are considered to be the best compromise between extending life and minimizing disorders for the areas and cultivars involved. Because of this variability,

Figure 3.1 Simple metabolism of fruit.

conditions described in this chapter should be used only as examples, and local guidance should always be sought.

3.2 History

There is evidence to suggest that the ancient Egyptians and Samarians in the second century BC stored portions of their crops in sealed limestone crypts to prolong storage life.

The first recorded experiments to control fruit ripening by changing the surrounding atmosphere were carried out in France by Jacques Berard in 1821. The results of his experiments won him the Grand Prix de Physique from the French Academy of Sciences, but failed to inspire any commercial application [2].

In the 1860s in Cleveland, Ohio a commercial storage operator called Nyce limited the oxygen available to his fruit in a sheet-metal-lined ice-cooled store. He realized the fruit generated 'carbonic acid', and he reported improved storage life and increased profits!

In 1918 Kidd and West started a thorough scientific investigation at the Food Investigation Organization in Cambridge. They had many problems to solve which required the cooperation of engineers and physicists to help overcome the practical problems of making gas-tight stores. Kidd and West later moved closer to the fruit-growing industry in Kent, and helped establish the Ditton Laboratory where they continued their pioneering work. In 1969 the Ditton Laboratory was incorporated into the East Malling Research Station, which has become the world's leading CA research institution. Pioneering work was carried on there in the 1960s with Fidler, Mann, and North making many improvements in storage techniques (Figures 3.2 and 3.3). This work has continued until the present day under the guidance of Sharples who, with the help of North, introduced the commercial use of ultra low oxygen (ULO) storage in the late 1970s.

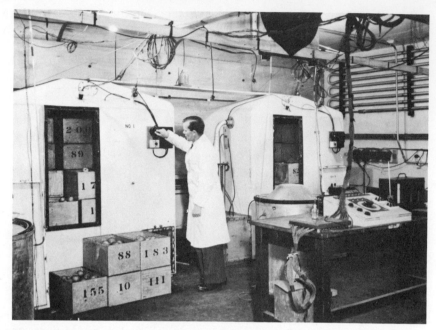

Figure 3.2 C.J. North carrying out early CA experiments at the Ditton Laboratory (Photograph courtesy of IHR, East Malling).

Figure 3.3 The Ditton Laboratory, East Malling Research Station (Photograph courtesy of IHR, East Malling).

Eaves working in Kentville, Nova Scotia introduced CA storage in North America in the early 1930s and was very innovative in developing new technologies and systems for CA.

Smock worked with Kidd and West in Cambridge in the 1930s and returned to his native USA where he published a paper in 1938 entitled *The possibilities of gas storage in the United States*. Smock continued to pioneer CA storage at his laboratory at Cornell University in Ithaca, New York, where he died in 1986.

The first commercial CA storage in the UK was carried out by Mr Spencer Mount in 1929. He successfully stored 30 tons of Bramley Seedling apples in 10% carbon dioxide at his farm in Canterbury, Kent. It is interesting to note that despite all the subsequent research, 10% carbon dioxide is still the recommended storage level for this cultivar.

At the outbreak of war in 1939 the storage capacity in the UK had grown to 30 000 tonnes. Expansion started again after the war with estimates of 70 000 tonnes in 1950, 100 000 tonnes in 1960, and 175 000 tonnes in 1966. The UK storage capacity in 1987 was approximately 300 000 tonnes. In 1965 there was 240 000 tonnes of CA storage in the USA, and this expanded to 1.8 million tonnes by 1987. CA storage started much later in continental Europe, with only 50 tonnes in 1959 in Italy and none in France until 1962. These countries have since expanded their CA storage rapidly.

Apples and pears remain the predominant crop stored in CA conditions, but other produce is also successfully stored. In 1977, work started at the Food Research Institute in Norwich (UK) by Geeson and others on the storage of cabbage, and this is now commercially practised for the year-round supply of Dutch White cabbage.

It has been shown that CA conditions can increase the storage life of green bananas. McGlasson and Wills in Australia found that at 20°C the storage life of an unripened banana could be extended from 16 days to 182 days in a CA atmosphere of 3% oxygen and 5% carbon dioxide [3].

Many other products respond to CA conditions, but its use is not currently widespread. Before a recommendation for storage conditions can be given, much research and testing over several seasons is necessary.

3.3 Controlled atmosphere conditions

As mentioned in the previous section, the first CA stores used a simple increase in the carbon dioxide levels. In this regime the carbon dioxide produced by respiration of the fruit is allowed to accumulate within the sealed store. The carbon dioxide level is measured and when it reaches a predetermined value the store is ventilated with air at a controlled rate to maintain the required concentration.

This storage system is still successfully and widely used in the UK for the storage of Bramley Seedling apples with a typical carbon dioxide level of 8%.

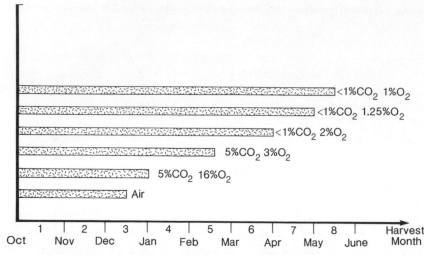

Figure 3.4 Storage life of Cox Orange Pippin apples in various atmospheres.

Because high carbon dioxide levels cause injury in many varieties, the next stage of CA storage involves the removal of carbon dioxide. This is removed from the atmosphere, and the oxygen level remains low as it is depleted by the fruits' respiration. The level of carbon dioxide is controlled by adjusting the amount of removal or 'scrubbing', and the oxygen is controlled by ventilating with air. A level of 5% carbon dioxide and 3% oxygen has been widely used for many years for the storage of most common varieties.

Low-oxygen storage with oxygen levels of 2% and carbon dioxide levels less than 1% have now superseded the 5:3 levels in many stores, and ULO storage using between 1% and 1.5% oxygen with lower than 1% carbon dioxide are used (with caution) in some of the most modern automatically controlled stores.

Figure 3.4 shows typical storage life of Cox Orange Pippin apples in different atmospheres, showing the increase in storage potential using lower oxygen conditions. As well as prolonging life the lower oxygen levels also improve firmness, texture, and the crispness of the marketed fruit. Some people consider there is a deterioration in flavour, and aroma especially at ULO conditions, but this can be improved by increasing the oxygen during the last few weeks of storage [4].

Table 3.1 gives a summary of the current (1988) recommendations for UK storage, and Table 3.2 gives some recommendations for Michigan, USA [5]. Again, as mentioned previously, these should be used only as guides and the detailed local recommendations must be consulted for commercial purposes.

3.4 Ethylene

Ethylene is produced by fruit as part of the ripening process and its presence accelerates ripening. CA storage reduces ethylene production, but it can still

Table 3.1 Typical UK storage conditions. For more details, see Sharples and Stow (1986).

Variety	Store to	Temp (°C)	CO_2	O_2	Notes
Apples					
Cox's Orange Pippin	Late March	3.5 to 4	1	2	
Cox's Orange Pippin	Late April	3.5 to 4.5	1	1.25	Auto only
Idared	May	3.5 to 4.5	5	3	
Idared	May	3.5 to 4.5	1	1.25	Auto only
Spartan	March	1.5 to 2	1	2	
Spartan	June	1.5 to 2	6	2	
Crispin	May	3.5 to 4	8		No scrub
Jonagold	April	1.5 to 2	8		No scrub
Jonagold	May	1.5 to 2	1	1.25	Auto only
Golden Delicious	April	1.5 to 3.5	5	3	
Red Delicious	March	0 to 1.0	5	3	
McIntosh	Mid March	3.5 to 4	1	2	
Bramley	June	4 to 4.5	8–10		No scrub
Bramley	June	4 to 4.5	6	2	
Pears					
Comice	March	−1 to −0.5			Air storage
Conference	April	−1 to −0.5			Air storage
Conference	May	−1 to −0.5	1	2	
Cabbage					
Dutch White	May–August	0 to 1	5	3	

build up to significant levels in CA rooms. It has been shown that reduction in these ethylene levels improves the storage life and reduces disorders [6]. Equipment is available for ethylene removal but because of its high cost this equipment is not in widespread commercial use.

Ethylene removal is also useful in the air storage of produce such as Kiwifruit and cut flowers, both of which are extremely sensitive to ethylene levels of around 0.1 ppm.

Table 3.2 Typical Michigan (USA) storage conditions. (MO = month only)

		Standard CA		Low O_2 CA	
		%		%	
Variety	Temperature °F(°C)	O_2	CO_2	O_2	CO_2
McIntosh	38 (3.3)	2–2.5	2.5 1st MO then 5.0	1.5	3.0
Jonathan	36 (2.2) 1st MO then 32 (0)	2–2.5	3–5.0	1.5	3.0
Empire	38 (3.8)	2–2.5	3–5.0	1.5	3.0
R. Delicious	32 (0)	2–2.5	3–5.0	1.5	3.0
G. Delicious	32 (0)	2–2.5	3–5.0	1.5	3.0
Idared	32 (0)	2–2.5	3–5.0	1.5	3.0
Law Rome	32 (0)	2–2.5	3–5.0	1.5	3.0
Rome	32 (0)	2–2.5	3–5.0	1.5	3.0
N. Spy	32 (0)	2–2.5	3–5.0	1.5	3.0
Mutsu	32 (0)	2–2.5	3–5.0	1.5	3.0

3.5 Alternative atomosphere conditions

There are other storage methods which, although not strictly CA, use similar techniques and equipment and so are mentioned here.

Modified atmosphere This primarily refers to transport systems such as the 'Tectrol' process whereby an atmosphere is generatd by external equipment and used to flush a transport container containing perishable produce. The container is then sealed and dispatched to its destination without any further maintenance or measurement of the atomosphere.

Modified atmosphere packing In recent years much work has been done on modified atmosphere packaging. In this system produce is packed in boxes or retail packs surrounded by a packaging material that controls the diffusion of oxygen and carbon dioxide, thus modifying the atmosphere within the pack [7]. The material can either have a controlled gas permeability or can be perforated in a predetermined way to produce a similar effect.

Ripening rooms Many tropical fruits are ripened under controlled conditions in specially built ripening rooms. In these rooms the temperature, humidity, and ethylene level are varied to produce the required ripening cycle. Under these conditions large amounts of carbon dioxide are produced by the fruit, and if these are not properly controlled damage to the produce and danger to the staff can occur. In some installations the carbon dioxide level is monitored and the ventilation is automatically controlled in order to keep this level acceptable.

3.6 Pre-storage chemical treatment

Most produce put into CA stores receives chemical treatment to reduce decay and some storage disorders. It is beyond the scope of this chapter to go into any detail but a summary will give a guide to what is required [8, 9].

The area is one of increasing complexity, as many countries are now limiting the amount of chemical residue on retailed fresh fruit and vegetables. Before proceeding with chemical treatment local advice must be sought. The usual method of applying these chemicals is called *drenching*, whereby bulk bins of fruit are saturated in water containing a mixture of the required chemicals.

Superficial scald or storage scald is an irregular area of burned or scalded appearance which forms on some varieties of apples during and after storage. This is controlled by applying an anti-oxidant immediately after harvest and before loading the fruit into the store. The chemicals used for this are diphenylamine or ethoxyquin.

To protect against various fungal rot problems in storage, collectively known as storage rot, the fruit can be dipped in a fungicide such as benzimidazole.

Figure 3.5 A typical CA installation of 10 × 120 tonne rooms at Maidstone, Kent, built by Harry Lawrence Ltd, Tonbridge.

Any chemical treatment should be part of an overall fruit management programme which includes the correct cultivation techniques and spraying programme. Great care should be taken to ensure that the fruit is in the best possible condition for long-term storage.

3.7 Storage rooms

The first consideration when planning a CA store is the size. A major consideration in determining size is the speed of loading and unloading. For optimum storage conditions a store should be completely filled with fruit of the same (or a compatible) variety in a matter of one or two days. When emptying a store the fruit should be graded and packed in 7–10 days after the store is opened [10, 11]. It is therefore preferable to have the total storage capacity divided into smaller units, but of course this is more expensive and economic considerations apply. In the UK an average store size would be about 100 tonnes, with variations between 50 and 200 tonnes (Figure 3.5). In North America the stores are larger, averaging 30 000 bu (600 tonnes), whilst in continental Europe the average size is about 200 tonnes.

CA stores are constructed in a similar manner to conventional cold stores described elsewhere in this book. A CA store must, however, be gas-tight, and this is achieved by careful attention to detail in design and workmanship. This is usually a job best left to experts, and many problems have been incurred by inexperienced contractors trying to build new stores or convert existing cold stores for CA use.

Almost all CA stores now being built in Europe are made from metal-faced insulating panels locked together with proprietary locking systems. Gas-tightness of the joints is usually increased by taping and coating with a flexible plastic paint. This process is also carried out on the floor–wall joints and wall–ceiling joints (Figure 3.6).

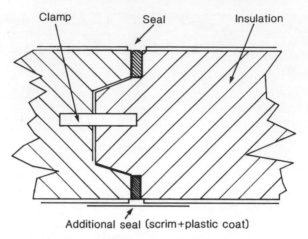

Figure 3.6 Joint between interlocking insulated panels.

Doors are a common leakage area, and they have to be of very substantial construction to allow sealing with a rubber gasket around the perimeter. These are sealed when the door is closed, with screw jacks equally spaced around the door. Some novel door seals are available from the Dutch firm Salco which are sealed once the door is in position by inflating a pneumatic seal with a simple foot pump. The doors on modern stores are almost always of the sliding variety and should be of sufficient height to allow efficient loading with FLTs.

Particular care has to be taken in gas-sealing all the internal fixings and entries for pipes and cables. Drains for the removal of condensate water have to be properly designed with U-traps to ensure that water can escape without breaking the room seal.

Each store should be fitted with inspection hatches to allow access to the fruit for routine examination during the storage period. It is common practice in Europe to include double or triple glazed windows usually adjacent to the store ceiling to allow visual inspection of the condition of the fruit and to check for the ice build-up on the refrigeration coils.

In North America panel-built stores are not so common. Here it is usual to construct the stores with either timber frame and plywood boarding, or with concrete blocks or tilt-up concrete walls. The store is then sealed and insulated with foamed-in-place urethane which is then coated with a fire retardant.

To protect the store structrue from damage due to excessive positive and negative pressures it is essential to install a pressure relief valve. This valve should limit the pressure on the store structure in either direction to 25 mm water. A simple system commonly used is a water trap. Figure 3.7 shows a

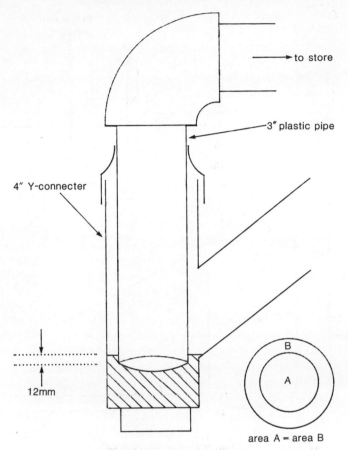

to store

3″ plastic pipe

4″ Y-connecter

12mm

B

A

area A = area B

Figure 3.7 CA room watertrap.

design from Michigan, USA which is easy to construct and reliable in operation. Care must however be taken to replenish the liquid regularly, and antifreeze should be added to prevent freezing. Spring-loaded commercial valves are available but these are not extensively used in either the UK or North America.

3.8 Store leakage specification

To check that a CA store is capable of being properly sealed it is essential for the store to be tested for leak tightness before it is loaded with produce. This is done by pressurizing the room and measuring the rate at which the pressure falls (Figure 3.8). The room is prepared for testing by checking that all doors, hatches, drains, valves, and pipes are close. A sensitive pressure gauge (manometer) is connected to the store; an inclined-tube water manometer is

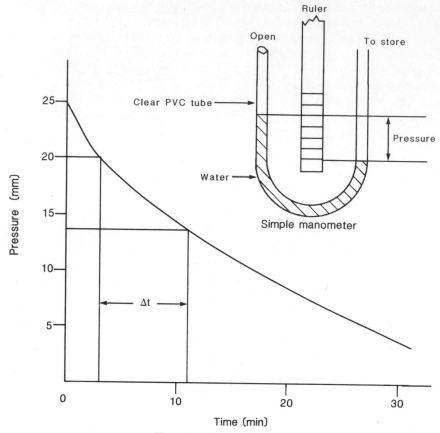

Figure 3.8 Store leak testing.

the best type to use. A dial-type bourdon tube pressure gauge must not be used, as this is nowhere near sensitive enough. After the manometer has been connected the store should be pressurized with a small air blower. A domestic vacuum cleaner can be used for this, but care is needed as these can produce enough pressure to cause structural damage even to very large stores. The store should be pressurized to 25 mm water gauge. When this is achieved the blower should be stopped and the air inlet to the store sealed. If 25 mm pressure cannot be achieved a large leak is indicated, and this requires rectification before continuing.

The rate at which the pressure falls is measured, and is an indication of the store leakage rate. In the UK the time taken for the pressure to fall from 20 mm water gauge to 13 mm is measured. In a store intended for storage at 2.5% oxygen and above, the minimum recommended time is 7 minutes. For stores running at 2% oxygen and lower, 10 minutes should be the minimum. It is not uncommon for well-constructed stores to take up to 30 minutes to lose 7 mm of water gauge pressure. North American operators define their tests slightly differently, and the time recorded is that required for the pressure to fall to half the starting value. For example, the time will be the same for a fall from 25.4 mm to 12.7 mm ($1''$ to $\frac{1}{2}''$) as for 12.7 mm to 6.4 mm ($\frac{1}{2}''$ to $\frac{1}{4}''$). The acceptable time for this is 30 minutes for all low-oxygen stores with all types of scrubber, but for 3% oxygen rooms 20 minutes is acceptable. Incidently the North American '20 min' room is equivalent to 12 minutes on the UK test, and the '30 min' room to 18 minutes.

3.9 Refrigeration

The design of the refrigeration system is covered in other chapters in this book, but a few notes on the requirements for CA stores are appropriate.

CA stores are by their nature long-term stores, and therefore product weight loss is a very important consideration in the design. The refrigeration system should be designed for a very small temperature difference between the produce and cooler surface, thus maintaining high humidity and minimizing the defrosting required. Secondary refrigeration systems using pumped glycol or brine are the most satisfactory, but many successful direct expansion systems with properly designed evaporators are in use. In recent years flooded freon or ammonia systems have become popular in the UK, and are regularly used in the USA.

To maximize the fruit storage potential the recommended store temperatures are specified to be close to the minimum before damage occurs. It is therefore important that close temperature control is achieved and that the temperature differences throughout the store are minimized.

Good circulation of air throughout the store is essential, and the recommended UK rate is in the range of 30–50 empty room volumes per hour. This can be reduced after temperature pull-down to half, or less, of the initial rate. A well-designed and operated CA store will have temperature differences within the store of less than 0.5°C. It is common practice to control the store temperature with an electronic thermostat or Coolstat with accuracies of better than 0.2°C and control differentials adjustable down to less than 0.5°C. Some installations in the UK have attempted to use standard proportional controllers, but in general these have proved unable to maintain the required accuracies.

These thermostats are used to control the air temperature in the store and the probe is normally mounted in a position away from the main air stream from the cooler to reflect the average store temperature. A convenient location for this is halfway up the rear wall of the store behind the cooler ducting. A second thermostat should always be used to prevent freezing of the fruit. This is particularly important in pear stores when an override thermostat controlling the minimum temperature of the air off the cooler should be fitted.

In order to minimize the refrigeration running time, the heat input to the store must be reduced to a minimum. A major source of heat input other than the respiration of fruit itself is the air circulation fan. These are sometimes switched off to reduce power, but great care must be taken to maintain even temperature distribution within the store. Some stores now use accurate thermometers to measure the differential temperature within the store and use this to control the fans either by variable speed fans or by on/off cycling. Both these methods have proved satisfactory and have given substantial power savings and as a result, less fruit weight loss. The reduction in fruit weight loss and improved quality are generally of a greater economic benefit than savings obtained by a reduction in electricity consumption.

If ammonia is used as a refrigerant and piped directly into the store evaporator, then some means of measuring and alarming for ammonia leakage within the store is recommended. A small ammonia leak can go unnoticed in a sealed CA store and can cause severe damage to the stored fruit.

3.10 Oxygen removal

As mentioned in section 3.3, CA storage generally requires a low oxygen level. The conditions within the store should be obtained as rapidly as possible to maximize the benefit of CA storage. It is recommended that the temperature and the gas atmospheres are achieved within 7–10 days of starting to load the store. To significantly reduce fruit development during storage, the CA conditions need to be established before the ripening processes have started. This is generally before the fruit starts producing a significant amount of ethylene, which is when the individual fruit's internal ethylene is less than 1 ppm.

Unless the stores are very leaky the removal of oxygen with external equipment is required only to achieve the initial conditions. For the remainder of the storage time the natural respiration of the fruit removes sufficient oxygen to overcome any small store leakage.

Natural respiration

It is common practice in the UK not to install machines for removal of oxygen. This is because much of the commonly stored fruit (i.e. Cox and Bramley apples) has a high respiration rate which reduces the oxygen concentration naturally to the required value in 7–10 days. Elsewhere in the world it is often normal practice to reduce the oxygen with some type of removal system.

Nitrogen flushing

Store oxygen levels can be reduced quickly by flushing with nitrogen. This can be purchased as liquid or gas and either stored on site or used directly from tankers. Used with care this can be an economic and safe way of reducing the store oxygen. The nitrogen itself may be expensive but there is very little capital or installation cost unless on-site storage of the gas is proposed.

The amount of nitrogen required can be calculated from the following approximate formula:

$$V = A \log_e (C_o/C)$$

where V is the volume of the gas to be injected, A is the store void space C_o is the initial oxygen concentration, and C is the oxygen concentration required. A store containing 100 tonnes of fruit typically would have a volume of $350\,m^3$ with a void air space of approximately 65%. This would therefore require $326\,m^3$ of nitrogen to reduce the oxygen in air (21%) to 5%, or $442\,m^3$ to reduce it to 3%. It is not necessary to reduce the oxygen completely to the required final level by flushing, as the respiration of the fruit can be used for the last one or two per cent.

The liquid nitrogen must be discharged slowly into the store to prevent the low temperature causing damage to the fruit. The fans must be running at full speed and the refrigeration turned off. The temperature in the store must be closely monitored and flushing stopped if it falls too low. In addition great care must be taken to prevent the pressure in the store exceeding 25 mm water or the structure could be severely damaged [12].

Propane burners

A propane burner uses the combustion of propane to convert the oxygen in the atmosphere to carbon dioxide and water. This principle is used in two ways in CA stores:

- An 'open flame' burner which reduces the oxygen in fresh atmospheric air: this is used to flush the store. A simplified equation for this is:

$$C_3H_8 + 5O_2 \rightarrow 3CO_2 + 4H_2O$$

- A propane burner with a catalyst enables combustion to continue with as little as 3% oxygen present, thus making it possible to recirculate the store atmosphere through the burner for greater efficiency.

A catalytic-type burner is the most common form of reducing store atmosphere levels but it is beginning to go out of favour in some areas, especially when low-oxygen storage is required (2% or less).

There have been problems with safety, due to the accumulation of propane and carbon monoxide which are produced when the burner is not correctly operated or adjusted. It is essential that an explosive gas detector is fitted to any system using this equipment. This type of burner also produces a large amount of carbon dioxide which has to be removed by the scrubber. Some ethylene is also produced, thus increasing the ethylene concentration in the store.

Ammonia crackers

In an ammonia cracker, anhydrous ammonia is reacted at high temperature with the recirculating store atmosphere to convert the oxygen to nitrogen and water with the hydrogen as an intermediate component:

$$4NH_3 \rightarrow 2N_2 + 6H_2$$

$$6H_2 + 3O_2 6H_2O$$

This has the advantage that no carbon dioxide or hydrocarbons are produced, but the disadvantage that the returning gas stream requires cooling and it is more expensive than the propane system.

Pressure swing

A pressure swing nitrogen generator uses the properties of a special carbon molecular sieve to separate the oxygen from atmospheric air. Two carbon beds are used, one adsorbing oxygen whilst the other is being vented and cleaned for the next cycle. This system requires the air to enter the sieve at a high pressure (8–10 bar) and the beds are cycled approximately every 60 seconds. The concentration of the oxygen in the outlet will depend on the flow rate and thus can be varied to obtain the optimum flow and oxygen level required. Care is needed with this system to ensure the air from the compressor is dry because water can clog up the bed and prevent proper operation.

Membranes

Membranes for separating oxygen and carbon dioxide have been used for many years, particularly in France. Modern developments in high-pressure membranes have recently been successfully introduced into the CA industry. Membranes employ the principal of selective permeation to separate gases. Each gas has a characteristic permeation rate that is a function of its ability to dissolve and diffuse through a membrane. This characteristic rate allows 'fast' gases such as oxygen, water vapour, and carbon dioxide to be separated from 'slow' gases such as nitrogen. This technique requires the high pressures of the pressure swing technology (8–10 bar) but is simpler in operation with no need for timing and change-over valves. A tube membrane is illustrated in Figure 3.9, but flat and coiled membranes are also available. Again the oxygen content varies with output flow rate and this can be adjusted to the optimum for the application.

Work is currently being carrried out to use this system in a recirculation mode which will increase its efficiency considerably. This will then approach the requirement of an 'ideal' machine which can remove both oxygen and carbon dioxide from the store atmosphere. This equipment is, however, relatively expensive and has higher running costs than alternative methods.

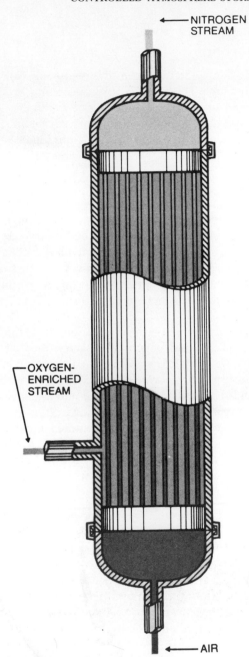

NITROGEN
STREAM

OXYGEN-
ENRICHED
STREAM

AIR

Figure 3.9 A hollow tube membrane (courtesy Permea Inc.).

3.11 Carbon dioxide removal

Lime scrubbers

The removal of carbon dioxide, or 'scrubbing' as it is commonly called can be achieved by using hydrated lime. This is still a very common practice and in many instances is preferred to mechanical scrubbers, especially for ULO stores where it is essential to limit the amount of oxygen entering the store. The lime that is used should be freshly hydrated high-calcium lime. Its reaction with carbon dioxide is

$$Ca(OH)_2 + CO_2 \rightarrow CaCO_3 + H_2O$$

The lime can be loaded in a 'lime box' external to the store and its effect on the store carbon dioxide level controlled by regulating the flow through the box. Alternatively, where there is a simple requirement for carbon dioxide levels in the store of less than 1%, the lime can be placed directly in the store with the fruit. In either method the lime should be in paper sacks. These are sufficiently porous to carbon dioxide and no additional puncturing is required. The bags should be well spaced to allow maximum exposure to the store atmosphere.

The amount of lime required depends on the respiration rate of the fruit, the length of storage time required, and which deployment method is used. Theoretically 1 kg of lime will adsorb 0.59 kg of carbon dioxide but in practice the limiting factor is the adsorption rate which decreases rapidly as the lime is exhausted. The recommended quantity of lime to place into the store for long-term Cox storage at 2% oxygen and 1% carbon dioxide is 5% of fruit weight. Other varieties, storage regimes, and periods could require a greater or lesser amount than this.

Once the lime is exhausted it is useful only for agricultural purposes and other means of disposal can sometimes be a problem.

Carbon scrubbers

Carbon dioxide can be adsorbed on the surface of activated carbon granules. Once the carbon is saturated no more adsorption takes place. However, if the carbon is then flushed with fresh air the carbon dioxide is removed and the carbon can be used again. This principle is widely used in carbon scrubbers. The store atmosphere can be blown through carbon beds with a low-pressure fan for a preset period of time (usually from 5 to 10 minutes). The store gas is then disconnected and the bed flushed with fresh air for a similar period. (Figure 3.10).

These scrubbers are commercially made with a single or a dual bed. The type and size chosen will depend on store size, fruit respiration rate and the level of oxygen and carbon dioxide desired in the store. An important factor in the design and setting up of a carbon scrubber is the amount of oxygen it

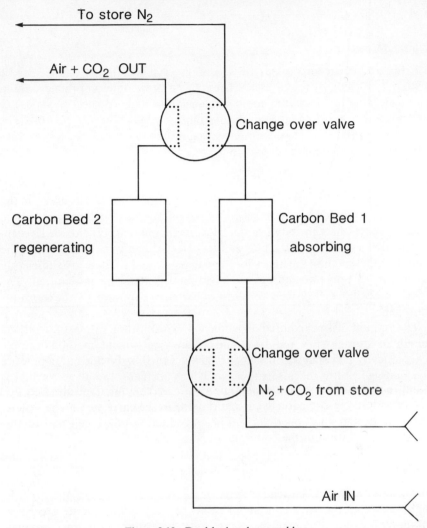

Figure 3.10 Dual bed carbon scrubber.

introduces to the store. This becomes critical at oxygen levels below about 3%, and the timing of valve operations to minimize this is most important.

The majority of carbon scrubbers will work satisfactorily at 5% carbon dioxide and 3% oxygen but only the very best scrubbers carefully maintained are adequate for ULO (1% oxygen, less than 1% carbon dioxide).

Water scrubbers

In older stores, particularly in North America, the use of water scrubbers to remove carbon dioxide was commonplace. This is now going out of practice,

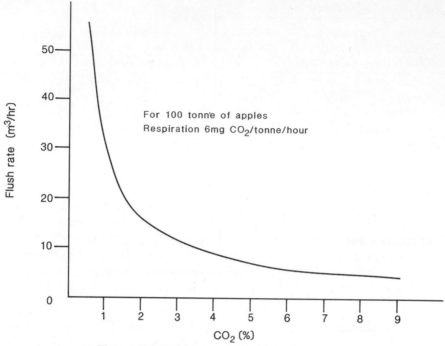

Figure 3.11   Flushing rate required for CO_2 removal.

mainly because water spray defrost is now no longer popular and also because it is difficult to remove enough carbon dioxide in the early part of the season when the fruit respiration is high. The introduction of oxygen due to the dissolved air present in the water also proved a problem.

Flushing

Nitrogen flushing can be used for removing carbon dioxide from the store. It might seem that the same equipment could be used as that for store oxygen pull-down (except of course those methods that produce carbon dioxide). However, on further examination it can be seen that a large amount of flushing gas is needed unless the carbon dioxide level required is relatively high. Because this technique flushes the carbon dioxide away, the volume required increases rapidly with the reduction in the carbon dioxide content, as shown in Figure 3.11. This graph can be used to determine the rate for other store sizes and respiration rates by multiplying by the appropriate factor [10, 13].

  Stores containing produce such as Bramley apples which do not require low oxygen but rely on high carbon dioxide levels are controlled by flushing with air. It should be noted that as this is flushing rather than oxygen replacement, considerably more air is required than for low-oxygen scrubbed stores. Allowance should be made in the design if this type of storage is envisaged.

3.12 Ethylene removal

It was mentioned in section 3.4 that ethylene removal from a store can have a beneficial effect on the fruit storage life and quality.

Ethylene can be adsorbed by potassium permanganate crystals [14]. These are successfully used for the transport of fresh produce, but prove uneconomic in long-term CA stores.

Commercial ethylene scrubbers have been made which remove the ethylene over precious-metal catalysts running at high temperatures. These are expensive in capital terms, and the energy needed to both heat the air and then to cool it again to acceptable levels proves difficult to justify. A novel solution is the swing therm concept [15] developed in Poland which reduces the energy consumption by passing the store gases through a porous heat exchange bed in alternate directions. Another commercial machine from Australia uses the same equipment as used for oxygen removal but the catalyst is run at a higher temperature when ethylene removal is required.

Ethylene can also be destroyed by subjecting it to ultraviolet radiation of certain wavelengths. This has been shown to operate in the laboratory but has not yet had any substantial commercial use [16].

3.13 Instrumentation

For successful CA storage of produce it is essential to have the correct instrumentation to measure the conditions within the storage room. More losses of stored products have occurred due to faulty measurement than any other cause. Equipment is needed to measure temperatures and gas concentration. This can be done with a variety of equipment varying from a simple portable apparatus to complete computer-controlled automatic systems. It must, however, always be stressed that any amount of computer or automation is irrelevant if the initial measuring method or accuracy is unsuitable for the task.

Temperature measurement

It is important that the temperature within the store is carefully monitored. If the temperature is too low, temperature breakdown or freezing injuries can occur; if it is too high, storage life is reduced. This is particularly important with the Cox and Bramley type of apple grown in northern Europe where the normal storage temperature is 3.5–4°C.

It has been common practice in the Uk for over 30 years to use a number of electrically operated temperature probes or 'drops' in each store to accurately monitor both the air temperature and the fruit temperature in various positions throughout the store.

In North America and many European countries the normal practice has been to use a single mercury-in-glass thermometer suspended in the store and visible through a window. This obviously gives satisfactory results with the easier-to-store varieties of fruit, but with recent trends to extend storage life, and to reduce power consumption by reducing fan operating time, it is prudent to install multiple probe equipment in modern CA stores.

The recommended minimum number of probes is three in a store holding 50 tonnes or less, increasing to four in a 100 tonne store and five or six in larger stores. One probe should be used for monitoring the temperature of the air freely circulating within the room and the remainder placed within the fruit at various locations to measure the actual fruit temperature. In a properly designed and loaded fruit store temperature differentials throughout the room should be less than 0.5°C, perhaps as little as 0.2°C.

The accuracy of measurement should be better than $\pm\,0.2°C$ and care must be taken when examining manufacturers' specification that the error of both instrument and probe are taken into consideration.

Platinum resistance thermometers (sometimes called PT100) can give accurate measurements of temperature and have been widely used especially in older stores. The accuracy of standard commercial probes to BS 1904 Grade 1 is $\pm\,0.25°C$, which is insufficient for this application, and therefore each probe requires calibration against an ice standard after installation. This will also take into consideration the connecting cables which because of the low basic resistance (100 Ohm) of the sensor, can cause significant errors. This can be reduced by the use of three- or four-wire connection systems. Another problem that occurs with thermometers using these sensors is errors due to changes in resistance of the probe selection switches.

Another type of sensor which has been used in storage rooms is the thermocouple. These have the advantage of being very inexpensive but the standard accuracy to BS 4937 type T class 1 is $\pm\,0.5°C$ which is insufficient unless individually calibrated. Thermocouples require special compensation cable and give extremely small voltage outputs (typically 40 microvolts per °C). Thermocouples are very good and useful sensors for high temperature applications but are unsuitable for use in modern cold stores.

The precision thermistor type of sensor is now widely used in fruit stores and it has many advantages. The basic resistance is high, allowing use with hundreds of metres of cable with negligible errors. The initial accuracy as purchased is typically 0.1°C and therefore individual on-site calibration is unnecessary although testing is always advised. There is sometimes doubt on the long-term stability of these devices which can be a problem on some of the lower cost units. The higher-quality probes, however, are extremely stable and practical measurements in many fruit stores have shown a total drift of less than 0.1°C over periods of 10–15 years.

Oxygen measurement

Oxygen is the most critical gas to measure in a modern low-oxygen CA store. Various types of oxygen analyser are available, but the most satisfactory is a sensor based on measuring the magnetic properties of oxygen. Oxygen is highly paramagnetic compared with all other common gases and this effect is used in a 'dumb-bell' measuring cell. The 'Servomex' oxygen analyser in various forms is in common use in the majority of low-oxygen fruit stores throughout the world [17]. Some care has to be taken with this type of analyser to ensure that the gas to be measured is properly clean and dry and the flow through the cell has to be adjusted to the correct level. It is necessary to calibrate the analyser against a known standard, and nitrogen or nitrogen/carbon dioxide mixture is used for setting the zero. Fresh air with a standard oxygen concentration of 21.0% is used to calibrate the full scale. For automatic oxygen control long-term analyser stability is essential and this is achieved by housing the analyser in a temperature-controlled cabinet.

For checking the operation of a fixed analyser system, it is always recommended that a portable oxygen analyser is used. A portable battery operated analyser suitable for this task is shown in Figure 3.12.

Alternative analysers include the fuel cell type which may be cheaper to purchase but do not have the accuracy of the paramagnetic and require frequent expensive replacement of the measuring sensor.

Analysers measuring oxygen with a zirconia cell can be excellent for very low levels of oxygen, but a very carefully controlled high-temperature furnace is required to achieve good accuracy. Again, for the levels of oxygen in a CA store the paramagnetic analyser remains superior to the zirconia cell.

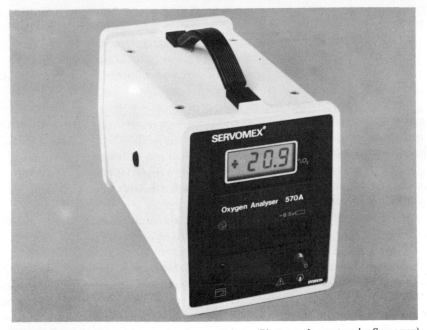

Figure 3.12 Portable paramagnetic oxygen analyser (Photograph courtesy by Servomex).

A chemical Orsat analyser is still commonly used in parts of North America and in southern Europe for fruit store analysis. The capital cost is low and the accuracy acceptable for higher oxygen regimes, but the readings are subject to chemical freshness and operator competence. The major drawback is the time needed to complete each reading and the dependence on a skilled operator. This equipment is not recommended for a modern CA installation and is unsuitable for low-oxygen storage.

Carbon dioxide measurement

The traditional method of measuring carbon dioxide in fruit stores before low-oxygen storage became commonplace was a thermal conductivity analyser. This type of equipment measures the temperature of a hot wire which changes as the thermal conductivity of the surrounding air increases with increasing carbon dioxide content. This is an acceptable method when the background gas remains stable (as in a carbon dioxide/air mixture) but when the relative concentrations of oxygen, carbon dioxide, and air are altered by scrubbing then the measured errors become unacceptable.

To ensure the measurement is specific to carbon dioxide, an analyser which measures the amount of infrared adsorption is used. In this type of analyser the infrared wavelength adsorbed by carbon dioxide is selected with special optical filters. This radiation is then directed through a simple cell containing the gas to be measured before being measured with a detector. The change in radiation is very small and therefore to improve stability the radiation is modulated either with a motor-driven chopper or by switching the source itself. The output from this measurement is inherently non-linear but this can be linearized electronically if required. This type of analyser is less sensitive to flow than the paramagnetic oxygen but care has still to be taken to ensure the sample is clean and free from water droplets.

Again the analyser will require calibration and it is usual to use air for zero and a nitrogen/carbon dioxide mixture of a known value for full scale adjustment. Atmospheric air contains approximately 400 ppm carbon dioxide but this does not cause significant zero errors in an analyser used for fruit storage. However, calibration air must be drawn from outside as room air can easily increase to significant carbon dioxide levels with people and fruit store ventilators in the area.

An Orsat chemical analyser can also be used for measuring carbon dioxide but the comments made about oxygen analysis still apply.

Also available is a simple chemical carbon dioxide indicator which is designed for flue gas analysis. In this equipment carbon dioxide is adsorbed in a chemical liquid and a graduated scale is used to measure the volume change. This equipment is not very accurate and cannot be generally recommended, but is used in some stores without serious problems.

Ethylene measurement

Ethylene in the concentrations required for fruit storage work is difficult to measure. For the higher ranges where ethylene is injected for ripening, a 'Drager' type analyser with glass tubes filled with chemicals that change colour can be used. The minimum detectable level with this equipment is typically 1 ppm.

For long-term storage where ethylene is to be removed, levels in the range 0.01–1 ppm require to be measured. The only practical method of measuring this is with a gas chromatograph which is an expensive machine requiring trained operators and practical only in the largest storage sites. A simple chromatograph from the USA designed specifically for fruit storage work is available, but again this requires careful use to obtain good results.

Humidity measurement

A certain level of humidity is frequently included in a specification for CA storage, often in the range of 90–98% which can be achieved by good refrigeration design. It is, however, very difficult to obtain a meaningful measurement of humidity at these ranges and usually the actual humidity level is much more stable than any measurement made. Most typical humidity sensors do not specify performance in the high humidity band, some do but quote a $\pm 10\%$ or more error; whilst some claim better performance, the author is unaware of any humidity sensor working on a long-term satisfactory basis in fruit stores.

The possible exceptions to this are a chilled mirror-type dewpoint meter and an aspirated type of wet-and-dry bulb thermometer. The need to measure relative humidity in fruit stores does not usually justify the initial expense or maintenance of this type of equipment.

3.14 Gas sampling systems

A very important aspect of gas monitoring within CA stores which must receive careful attention is the sampling of the storage atmosphere. The analysers must be presented with a sample to measure that is clean, free of water droplets, and representive of the gas within the store. The time taken to obtain this sample should be as short as possible to prevent wasted time waiting for the sample.

There is no evidence to suggest that there is any stratification of gas layers within a store even with the refrigeration circulation fan on intermittent or variable duty. The sample should be taken away from any entries or exits of atmosphere control equipment. A good place is halfway up the back wall of the store near the cooler. A filter can be usefully fitted to the end of the sampling tube to prevent rubbish entering the system. If the store is operating below 0°C it is better not to fit a filter, however, as condensate running back down the line can freeze and block it.

The tubing connecting the store sample point to the measuring position should be at least 6 mm inside diameter to help prevent blocking with water. Condensation can occur in sample lines when the ambient is less than the storage temperature. The lines should always be sloped either towards the store or to the analysing station to ensure that the condensate drains away. If the lines are exposed and very cold weather is expected trace heating should be considered to prevent the lines freezing. This is particularly important on automatic control systems in which the sample is taken much more frequently than manual systems and the store condition is dependent on the automatic control.

Air leaking into sample lines is the most common cause of disaster with CA stores. This can be prevented by taking care in the installation and testing. The tubing material should be either copper or a heavy-duty solid plastic. Many of the lightweight clear flexible tubes turn brittle and crack over a period of time. The tube should be installed as a single length if at all possible. Copper tubes should be brazed in preference to using tube connectors. The junction into the store should be without joints, using electrical-type glands rather than conventional pipe fittings. When completed, *all* sampling lines should be vacuum leak tested and this should be repeated on a regular yearly basis.

The sampling lines should be terminated directly to a valve manifold which must be leak-tight to prevent cross-contamination of the samples.

A pump is required to pull the sample from the store and to provide sufficient pressure to supply the analysers. It is important that the pump does not also pull in ambient air and that the output is oil-free. A diaphragm pump is the most suitable but regular checks should be made on the diaphragm. Incorrect samples due to splits have been known to result in serious damage being caused to fruit in the store.

A filter and catchpot is then needed to ensure the sample is clean and free of water before connecting to the analysers.

3.15 Store gas control

Oxygen control

To control the oxygen level in the store, air is allowed to enter in a controlled manner. It is usual to construct a store with fresh and foul air pipes, as in Figure 3.13 and these are fitted with manually operated valves. These valves need to be very carefully adjusted regularly, usually twice daily, based on the oxygen readings obtained from the manual analysers.

The amount of air required depends very much on store leakage rate and fruit respiration, but as a guide 100 tonnes of apples respiring at 6 g of carbon dioxide per tonne per hour require about 1.5 m^3 of air per hour to maintain the

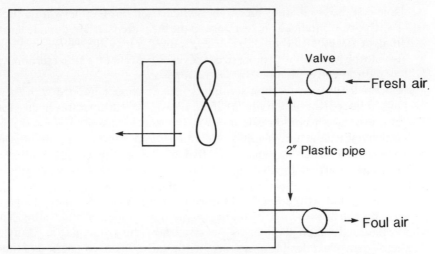

Figure 3.13   Store ventilation.

required oxygen level. As described earlier, the manual control of oxygen is quite tedious and requires constant adjustment. Automatic control of oxygen is desirable at lower oxygen levels and is considered essential at ULO conditions. As well as making life much easier for the operator, automatic oxygen control stabilizes the oxygen level which gives measurable increases in the quality of the stored fruit. Many CA stores have been automated by adding a simple solenoid valve to the fresh air line. An automatic controller then measures the oxygen in the store on a regular basis and when it is found to be low the fresh air valve is automatically opened for a period of time. Because of the very slow response time of a fruit store this type of control can easily maintain the level required with a stability of better than 0.1% oxygen.

It is sometimes necessary to use external fan assistance to ventilate the store and this particularly applies when the store circulation fans are controlled to minimize the energy usage.

Carbon dioxide control

The control of carbon dioxide depends on the method of removal chosen, as described in section 3.11

- *Flushing* If carbon dioxide is controlled by flushing with either air or nitrogen the flushing gas is switched with valve or solenoid when the carbon dioxide increases above a preset level. This is done either manually or automatically.
- *Lime scrubbers* Certain storage regimes require lowest possible carbon dioxide level; to achieve this lime is often put directly into the store, with the result that no control is possible. Where lime is placed in an external

lime box the circulating fan can be switched on or off either with a time switch or automatically in response to the measured carbon dioxide level. If higher levels of carbon dioxide are required it may also be necessary to isolate the lime box with motorized valves to prevent unwanted scrubbing by natural convection.

- *Carbon scrubbers*   Carbon scrubbers usually come equipped with their own time control system which can be adjusted by the operator to give the best average carbon dioxide levels in the stores. This type of control is usually adequate and is used in the majority of instances. Equipment is now available for automatic control of the scrubbers based on the measured carbon dioxide levels. This can have some advantages in optimizing the scrubber operation and minimizing power consumption, but it is not nearly as useful or important as automatically controlling the oxygen.

3.16   Automatic controllers

It is common practice in the UK and northern Europe to automatically control low-oxygen CA stores. These controllers can be based either on microcomputers or analogue electronics [18,19].

The gas levels in each store are periodically measured, and the value compared with the required level which is preset into the machine. If the oxygen level is lower than required the store is automatically vented, and if carbon dioxide control is also fitted the scrubbers automatically operate when the carbon dioxide level is high.

In the more sophisticated controllers temperature is also measured, and each store can be individually programmed to the user's requirements. The computer-based controllers also produce a regular printout for record keeping as well as many alarm functions. Some controllers can even automatically calibrate the gas analysers, thus further reducing the manual effort required. Automation should not, however, be an excuse for neglecting the stores, and each installation should be checked daily for correct operation. It is strongly recommended with any automatic or central measuring system that the gas levels are regularly (say weekly) checked with a portable analyser connected directly to the store [19]. This then checks for any error due to sampling line faults and analyser calibrations.

Figure 3.14 shows an Oxystat automatic fruit store control system.

3.17   Regulations

In the UK there are no government regulations that apply specifically to CA fruit.

In North America apples bearing the CA label are subject to regulations which vary from state to state. These regulations are old and badly in need of

Figure 3.14 Oxystat computerized fruit store control system (Photograph courtesy of D. Bishop Instruments).

revision as they bear very little relationship to conditions currently required for good long-term storage. Typical regulations for fruit bearing a CA label would be storage for 90 days or longer at 5% or lower oxygen (60 days for Jonathan apples). 5% must be achieved within 30 days of loading and various rules apply about notification, equipment, and licensing.

3.18 Safety

Low oxygen

CA stores by their very nature are dangerous because of the low oxygen content of the atmosphere. There are regularly fatal and near-fatal accidents in CA rooms [20]. No one should ever enter or even place their head inside an operating CA room. People have lost consciousness, fallen into the room, and died just a few feet inside the doorway. *Never* enter alone, and never open a door, hatch, or window without having at least one other person familiar with the hazard near. A store should be clearly labelled with caution and danger signs, and doors should be locked once CA conditions are started. The symptoms of asphyxia given below should be familiar to all personnel operating CA stores [21].

- *21% oxygen* Breathing, all functions normal
- *17% oxygen* Candle is extinguished
- *12–16% oxygen* Breathing increased and pluse rate accelerated. Ability to maintain attention and to think clearly is diminished, but can be restored with effort. Muscular coordination for finer skilled movement is somewhat disturbed.
- *10–14% oxygen* Conciousness continues, but judgement becomes faulty. Severe injuries (burns, bruises, broken bones) may cause no pain. Muscular efforts lead to rapid fatigue, may permanently injure the heart, and induce fainting.
- *6–10% oxygen* Nausea and vomiting may appear. Legs give way, person cannot walk, stand, or even crawl. This is often the first and only warning and it comes too late. The person may realize he is dying, but he does not greatly care. It is all quite painless.
- *Less than 6% oxygen* Loss of consciousness in 30–45 seconds if resting, sooner if active. Breathing in gasps, followed by convulsive movements, then breathing stops. Heart may continue beating a few minutes, then it stops.

Remember: CA storage contains less than 5% oxygen

High carbon dioxide

In some circumstances such as ripening rooms or cold rooms containing respiring fruit the carbon dioxide produced by the fruit can, if insufficiently ventilated, build up to dangerous levels. For example, levels of carbon dioxide exceeding 5% are regularly obtained in banana ripening rooms.

The limits of carbon dioxide in rooms for human occupation are (HSE 1986):

- *Continuous occupation*   0.5%
- *10 minute exposure*   1.5%

To ensure these levels are not exceeded holding rooms should be continuously ventilated. If this is not practical the carbon dioxide level should be monitored and the room ventilated when the carbon dioxide level is excessive.

Hydrocarbons

If a propane or similar type of fuelled burner is used for reducing the oxygen, incomplete combustion can cause a build up of hydrocarbons within the room.

If the lower explosive limit (LEL) is reached, a small electrical spark or source of ignition can cause an explosion. Some stores have been completely demolished in this way, and it is considered a serious enough hazard for many operators to replace the burning type of oxygen removal equipment with the much safer nitrogen generators.

If burners are used great care must be taken to ensure they are operated within the manufacturer's recommendations. It is also now considered essential to install an LEL measuring instrument to sample the store atmosphere for dangerous levels of hydrocarbons.

Fork lift trucks

Because CA rooms are of sealed construction the operation of fossil-fuelled FLTs within the room when loading cause carbon monoxide to build up to dangerous levels [22]. The physiological effect of carbon monoxide on humans is as follows:

- 50 ppm (0.005%)   Safe for continuous exposure
- 100 ppm (0.01%)   No perceptible effect
- 200 ppm (0.02%)   Slight effect after six hours
- 400 ppm (0.04%)   Headache after three hours
- 900 ppm (0.09%)   Headache and nausea after one hour
- 1500 ppm (0.15%) Death after one hour

50 ppm is the OSHA established maximum acceptable level for carbon monoxide. The National Institute for Occupational Safety and Health has recommended lowering the level to 35 ppm.

A propane lift truck producing 6% carbon monoxide in the exhaust is not uncommon. Careful tuning can reduce this to less than 1% but this is still dangerous and should not be used in unventilated areas. A properly working catalytic converter on the exhaust can remove 90–99% of the carbon monoxide, making it safer to used in enclosed storage.

The use of electric FLTs in CA rooms is much preferred from a safety standpoint, but if propane fuelled trucks are used they *must* be fitted with a functioning catalytic converter. Petrol fuelled trucks should not be used within a CA room.

3.19 Store operating practice

The following is a summary of the necessary actions required for successful CA storage. If there is any doubt on what action is required further advice should be sought from the many experts.

Before start of season

- Check all gas sampling lines for leaks
- Check all temperature probes for accuracy

- Check all analysers are operating correctly
- Check stores are leak-tight and pass the recommended pressure
- tests
- Check scrubbers and nitrogen generators for leakage and correct operation

Before loading
- Check chemical drenches are made up to the correct concentrations
- Check that the store is loaded uniformly and that the store is full
- Check the temperature probes are in the fruit bins and evenly placed throughout the store
- Load and obtain conditions as rapidly as possible

During loading
- Check the fruit mineral analysis is acceptable (especially for long-term storage)

After sealing
- Measure and *record* gas and temperatures twice a day immediately after sealing
- Check analysers *daily* with fresh air
- Calibrate analysers with test gas every *two days*
- Take independent gas readings with a portable analyser directly from the store at weekly intervals. If difference is greater than 0.2% put fans on high speed for two hours, recalibrate analysers and check again. If error still present ask for service
- Inspect samples of fruit from the store on a regular basis
- If store oxygen is less than 1.8%, check fruit samples for alcohol at monthly intervals

Despite the many precautions and care required, CA storage is used very successfully throughout the world to allow year-round availability of quality fruit and vegetables.

3.20 Future developments

CA storage remains a healthy industry, with future expansion and growth likely to occur in the following three areas.

Other products

Whilst apples and pears will probably continue to dominate the CA industry, there is continued interest in extending the storage life of other fruits and vegetables. Cabbage storage has expanded in recent years and this can be expected to extend to other leaf crops. It has been shown that soft fruits respond well to CA and although storage periods are much shorter useful extensions to the marketing seasons can be obtained.

To determine the optimum storage conditions for each crop and variety requires a considerable amount of research work. With the continued reduction in government-sponsored research in near-market areas it will fall increasingly on the growing and distribution industry to fund the necessary experiments.

Reduced use of chemicals

The continued consumer pressure to reduce the amount of chemicals used on fresh produce will cause changes in the CA industry. Some countries are already banning the sale of fruit treated with chemicals commonly used to reduce storage disorders (e.g. diphenylamine, ethoxyquin, captan, etc). As an alternative to some of these chemicals, changes to CA regimes can give beneficial results. For example, in the UK it has been shown in Bramley apples that scald can be reduced to acceptable levels without chemical treatment by changing the CA conditions from 8% carbon dioxide and 13% oxygen to 2% oxygen and 6% carbon dioxide [4].

Work is likely to continue in this direction, with the probability that it will become more commonplace to use lower oxygen levels than at present. It may also be more common to use ethylene removal as an additional CA requirement.

Transport

There is a world-wide movement towards the CA transport of perishable products. This will expand in two areas. Traditional top fruit exporters shipping fruit to the opposite hemisphere are seeking ways to expand their market share. This can be done by shipping under CA conditions which can both improve quality and extend the marketing period.

The other area of expansion is in the shipping of exotic fruits from the tropics to the temperate consuming nations. Currently the fruit, which has a typical shelf life of 7–14 days, is exported by air with all the resultant costs. CA conditions would extend the storage life sufficiently to enable shipment by sea. This trade is likely to be container-based but considerable work has yet to be done to increase the reliability of CA containers used for regular shipments of products. Another major problem for this trade is the correct selection and handling of fruit at harvest to ensure suitability for storage and transport.

Acknowledgements

The author acknowledges, with thanks, the help of D.J. Chappell, M.Inst.R., Fruit Store Consultant, Professor D.R. Dilley, Department of Horticulture, Michigan State University, and

Dr R.O. Sharples, Fruit Storage and Market Quality Department, IHR, East Malling, Kent, in reviewing this chapter.

References and further readings

[1]   Fidler, J.C., Wilkinson, B.G., Edney, K.L. and Sharples, R.O. *The Biology of Apple and Pear Storage*. Commonwealth Agricultural Bureaux, London.

[2]   Dalrymple, D.G. (1967) The development of controlled atmosphere storage of fruit. Publ. Div. Mktg Utiln Sci., U.S. Dept. of Agriculture.

[3]   McGlasson, W.D. and Wills, R.B.H. (1971) Effect of oxygen and carbon dioxide on respiration, storage life and organic acids of green bananas. *Aust. J. Biol. Sci.* 24, 1103.

[4]   Sharples, R.O. and Stow, J.R. (1986) *East Malling Research Station Annual Report for 1985*, 165–170.

[5]   Dilley, D.R. (1988) Recommended storage conditions for apples post harvest. Dept of Horticulture, Michigan State University, East Lansing.

[6]   Dover, C. (1983) The control of superficial scald of Bramley's Sedling apples by removal of ethylene from the storage atmosphere. *Proc. 16th Int. Conf. Refrigeration*, pp. 165–170.

[7]   Geeson, J.D. (1987) Prolonged life for fresh fruit and vegetables. *Food Technology International—Europe*. Sterling Publications.

[8]   Smock, R.M., Rosenberger, D.A., and Blanpied, G.D. (1984) Post harvest chemical treatments for reduction of storage scald, rots, and senescent breakdown of apples in storage. Cornell University, Ithaca, NY.

[9]   Johnson, D.S. (1979) Post harvest chemical treatments for controlling apple storage disorders. *East Malling Research Station Annual Report for 1978*, 217–219.

[10]   MAFF, *Refrigerated Storage of Fruit and Vegetables*, MAFF Reference Book 324, HMSO, London.

[11]   Bartsch, J. and Blanpied, G.D. Refrigeration and controlled atmosphere storage for horticultural crops. NRAES-22, Northeast Regional Agricultural Engineering Service, Cornell University, Ithaca, NY.

[12]   Bartsch, J. Creating a low oxygen atmosphere with liquid nitrogen. EF 9, Dept of Agricultural Engineering, Cornell University, Ithaca, NY.

[13]   USDA, *The Commercial Storage of Fruits, Vegetables and Florist and Nursery Stocks*. Agricultural Handbook 66, U.S. Dept of Agriculture.

[14]   Lister, P.D., Lawrence, R.A., Blanpied, G.D. and McRae, K.B. (1985) Laboratory evaluation of potassium permanganate for ethylene removal from CA apple storages. *Trans. ASAE* 28, 331–334.

[15]   Wojciechowski, J. and Haber, H. (1982) Swingtherm—a new economic process for the catalytic burning flue gases. *App. Catal.* 4, 275–280.

[16]   Shorter, A.J. and Scott, K.J. (1986) Removal of ethylene for air and low oxygen atmospheres with UV radiation. 19 (2), 176–179, CSIRO, Australia.

[17]   Tipping, F.T. (1974) The determination of oxygen in gases. Presented at the Institute of Measurement & Control Spring School of Process Analytical Instrumentation, March 1974.

[18]   Bishop, D.J. (1981) The application of micro-processors to controlled atmosphere fruit storage. *Agricultural Engineering* (1981), 51–54.

[19]   Lougheed, E.C. and Bishop, D.J. (1984) Electronic monitoring and control of carbon dioxide and oxygen in controlled atmospheres. *Acta Horticulturala* 157, 51–56.

[20]   Bartsch, J. CA Storage. EF 8, Dept of Agricultural Engineering, Cornell University, Ithaca, NY.

[21]   Henterson, Y. and Haggard, H.W. *Noxious Gases and the Principles of Respiration Influencing their Action*. Reinhold, New York.

[22]   Dilley, D.R. Precautions about carbon monoxide in CA storage rooms and packing houses. Dept of Horticulture, Michigan State University, East Lansing.

4 Energy conservation

N.P. DANIELS and A.O. PAGE

4.1 Introduction

This chapter considers the methods of energy conservation available to the cold store operator.

Many cold stores were built in an age when electricity was relatively cheap and the main criterion for purchasing a cold store and its associated cooling equipment was minimum capital cost—running cost was relatively unimportant. With the rising cost of electricity, the cost of operating cold stores has risen and operators have had to address the problem of reducing the running costs of the plant.

Operators might consider the purchase of new equipment or a new store. The design stage is the most cost-effective time to implement any energy conservation methods and to select the plant that will have the least running cost. This chapter discusses the various plant configurations that should be considered to minimize the running cost of the cold store.

4.2 How is cold produced?

The method chosen to produce cold depends on a number of factors, including:

- The temperature required
- The product being cooled
- The source of energy for the production of cold

Cold can be produced by

- Vapour compression cycle
- Absorption cycle
- Air cycle
- Total loss refrigerants, e.g. liquid nitrogen

The vapour compression cycle is by far the most common method of producing cold. These systems are described in a later chapter, and, for the purposes of this book, the other systems may be ignored.

4.3 Equipment considerations

In the following sections of this chapter, the various types of equipment used in the production of cold for the end user, are considered. In each case, the good practices that should be adopted when using the equipment are discussed, and particular reference is made to methods of ensuring that the minimum amount of energy is expended in producing the cooling and operating the cold store.

4.4 Finnned tube evaporators

In this section, the evaporators that will be discussed in detail are the direct air coolers with a fluid on one side and air on the other, and the older wall-mounted tube evaporators relying on convective heat distribution to the air. Liquid coolers are omitted from the detailed discussion in this chapter, but they are mentioned in general terms. The air-side performance of the heat exchangers is first discussed, and then the best way to engineer the refrigerant side.

Air-side design

The evaporators used in cold stores and for food freezing are normally made of finned tubes arranged in banks to get as much surface into a given space as possible. In some instances plain tubes are used, especially where heavy frosting might occur, but in cold store applications this is unlikely to be the case. A disadvantage of plain tubes is that a large number of tubes is required to provide the necessary surface for heat transfer between the fluid and the air.

Several types of fin design are available for air cooling coils. Some of them are shown in Figure 4.1. In general terms the designer's aim is to get as much surface attached to the tube as possible, to improve the heat transfer as much as possible. When attaching the fins to the tubes, great care must be taken to

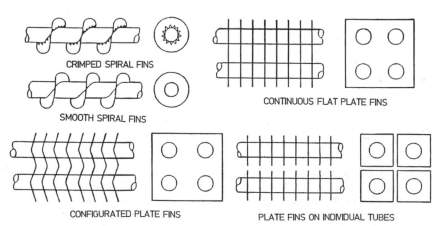

CRIMPED SPIRAL FINS

SMOOTH SPIRAL FINS

CONTINUOUS FLAT PLATE FINS

CONFIGURATED PLATE FINS

PLATE FINS ON INDIVIDUAL TUBES

Figure 4.1 Types of fin for air cooling coils.

ensure that intimate contact is permanently maintained, to ensure unimpeded heat transfer. The tubes in the heat exchanger can either be mounted in-line or staggered. The latter arrangement increases the turbulence and therefore the heat transfer. The fins can be either plain or corrugated, the latter giving better heat transfer. The design and surface arrangement have a great effect on the air-film heat transfer coefficient and also on the air-side pressure drop. The tubes are typically plain inside, though spiral inserts can be used to promote turbulence and improve the heat transfer of the fluid inside the tubes; a disadvantage with inserts is the increased pressure drop in the tubes with consequent increase in running costs.

It is one of the main aims of the operator of a cold store to minimize the running costs of the store. One way to achieve this is to operate the refrigeration plant at as high an evaporating temperature as possible. With a fixed air temperature in the store this can be done in a number of ways, but each method can have a cost penalty either at the installation stage or in the running cost of the plant. The operation and design of the air side of the evaporator will be considered first.

To raise the evaporating temperature to as near room temperature as possible requires increasing the heat transfer rate between the cooling medium and the air. This can be done in one of two ways:

- Increasing the surface area of the coil by installing a large number of fins. To achieve this the fins can be installed as close together as possible to reduce the size of the coil. This has implications in the cost of the coil and also in the size of the space required to locate the coils in the store. The closer together the fins the greater the pressure drop across the coil, which in turn will increase the running costs. At the design stage of the cold store it must be decided that the optimum sized cooling coil will be installed for minimum operating costs. Fins close together will result in the need to defrost the coils more frequently as the ice will bridge the fins quickly. This will result in poor temperature control in the store and higher running costs.
- Increasing the air velocity over the coil. This will also increase the pressure drop across the coil and the running costs.

A trade-off between the physical size of the coil and the running costs must be made to optimize the size of the coil.

As far as the refrigeration plant is concerned the operation of the coils is economic if the temperature difference between the refrigerant evaporating temperature and the air temperature is as small as possible so that the evaporating temperature can be as high as possible. The added advantage of operating in this fashion is that the humidity level in the store will be maintained at a higher level.

It is common in modern cold stores to install the cooling coils at one end of the cold store and to have high-velocity air blowing over the coils to 'throw' the air to the back of the store. Compared to the use of ducting in the cold store this is inefficient as the fan power required to 'throw' the cold air to the back of large stores is very high. The disadvantage of using ductwork to duct the air to the back of the store is that it occupies valuable storage space and reduces the available head height for racking.

Defrosting

Any unit cooler operating at or below 0°C is liable to the accumulation of frost or ice on the cooling coil. While a thin layer of frost does improve the heat transfer, if the layer is allowed to increase then the heat transfer will fall off. The result of this is that the evaporating temperature will decrease and compound the problem. In addition the pressure drop across the coil will increase, which can lead to an increase in fan running current.

It is therefore essential that the level of frost accumulation should not be allowed to increase to a level where it affects the performance of the plant and the operation of the cold store.

Defrosting can be initiated by a number of methods

- *Timed defrost* Based on a pre-set time, the initiation of the defrost is at regular intervals. This can operate quite well in a store where there is a constant vapour gain by the store. This is seldom the case, however, and this method can lead to unnecessarily high running costs due to defrost cycles being carried out when not required. Typically the timer will be set to ensure frost-free operation during the daytime when the vapour load on the store will be high due to the constant operation of the doors.
- *Ice sensors* There are a number of ways in which the presence of ice can be detected. Once the level of ice detected has reached the set point the defrost cycle will be initiated. The types of sensors available are:

  –air-side pressure drop. This can give misleading results due to the air bypassing the coil through the drain pan
  –active infrared ice detector. These operate either on the burglar alarm principle or on the variation in reflection as the ice layer builds up
  –fan current rise. This operates on the same principle as the air-side pressure drop and monitors the fan current which will rise when the ice forms on the fins

Other methods include air temperature difference measurement, which will increase with reduced air flow, and monitoring compressor suction pressure, which will decrease with increase ice fouling on the evaporator.

Other methods may be supplied by certain contractors. Each will have its advantages and benefits, but in general terms those that measure the build-up of ice on the coil will probably give the lower running costs for the plant.

Once defrost is initiated, how long should it last? The simple answer to that question is, 'until there is no ice left on the coils'! The measurement of whether ice is left on the coils or not is difficult. Termination of defrost depends on the type of defrost system. With some types of hot gas defrost it can be done on refrigerant pressure, but otherwise it is probably best done by a timer which will be set from experience. With the defrost initiation sensors which measure a build-up of frost it is likely that the amount of frost on the coils will be the same at the start of the defrost cycle. Making this assumption the timed defrost cycle will, on the whole, give adequate results.

There are a number of methods of defrosting the coils, and the chosen method will depend somewhat on the equipment supplier. The simplest way of defrosting coils, which can be used for air temperatures above 5°C, is to turn the refrigerating plant off and allow the fans to continue running. The room air temperature will be higher than the frost temperature and the ice will melt. This method is slow, however, and it is likely that much of the water will return into the air in the store.

There are three main methods of defrosting coils.

- *Spray defrosting*   This is achieved by spraying water or brine onto the coil surface. Water alone has successfully been used down to −40°C, though at these temperatures it is necessary to have adequate heating for the drain pan, drains, and supply lines. This method is, however, very messy.
- *Electric defrosting*   Electric heating elements are installed into the coil block. The effectiveness of this method depends on the location of the heating elements. Electric defrosting is simple, can be effective and rapid, but is relatively expensive to operate. This method of defrosting can be used at very low temperatures.
- *Hot gas defrosting*   This method utilizes the heat in the discharge gases from the compressor to defrost the ice on the outside of the tubes. There are several ways this can be done, some of which are considerably more energy efficient than others. In general this method can be as quick as the water sprays but has the extra advantage that it can be used at very low temperatures. A further advantage is that the injection of hot gas into the evaporator facilitates the return of oil to the compressor and acts as a method of cleansing the evaporator of oil during the defrost cycle.

Where secondary refrigerants are used as the fluid for cooling the air stream it is possible to utilize spray and electric defrosting as described above. In addition heaters can be installed in the secondary refrigerant circuit which warm up the fluid for defrosting the coil. This type of system provides heat from within the coil and is as rapid as hot gas defrosting. Where the speed of the defrosting cycle is critical a combination of two or more of the above methods can be used.

During defrost it is essential that:

- The fans are stopped

- The refrigeration plant serving the coil is stopped or isolated from the circuit

To ensure that all the water from the defrost cycle drains away properly, the drain pan and drain lines should be heated. In addition the drain lines should be adequately trapped, preferably outside the cold area to prevent freezing, and pitched to ensure rapid draining. The heating elements of the pan and drains need only operate for the duration of the defrost cycle.

Once the defrost cycle has been completed and the refrigeration to the coil reinstated it is important that the fan is not started until the coil has reached its operating temperature. This will minimize the heat gain by the store and will also ensure that any water droplets on the coil face are not blown into the store.

4.5 Wall tube evaporators

As an alternative to finned tube evaporators with forced convection, i.e. a fan, for cooling the stores, cold stores were constructed with banks of tubes mounted on the vertical walls of the store. These were the evaporator tubes into which high-pressure liquid refrigerant was passed through manually operated expansion valves. To cool the store a large surface area of tubes was required and as a result the capital cost of the installation was high. In addition the refrigerant charge was very large, adding to the cost of the installation. The advantage, however, was that the approach temperature to the store air temperature was very low and as a result the air was not dehumidified as much as in modern stores. The running costs of these stores is less, due to the higher evaporating temperature in the heat exchangers. These stores went out of vogue as a result of the large installation costs and quantity of refrigerant involved, and it is unlikely that they will return.

4.6 Refrigerant circuits in evaporators

There are three basic refrigeration circuits which can be used to provide the necessary cooling to a cold store. These are:

- Direct expansion, which utilizes a thermostatic expansion valve, and is used extensively on air conditioning plant, etc.
- Gravity recirculation
- Pumped recirculation

Of the three the direct expansion system is the least energy efficient but is generally cheaper and simpler to install. The benefits of not choosing a direct expansion system outweigh the extra capital costs incurred. This is discussed in greater depth in the following sections:

Direct expansion

Of the various types of evaporator designs available, this is by far the most costly in energy consumption. A typical evaporator operating in this manner is shown in Figure 4.2. There are two reasons why this system is poor in its efficient use of energy and capital:

- The expansion valve requires a superheat at the evaporator outlet so that it can control the liquid flow rate into the evaporator. To achieve the superheat it is necessary for a portion of the evaporator surface to be dry. This part of the evaporator is contributing little towards the cooling of the airstream as the heat transfer between gaseous refrigerant and the tube is very poor compared to the boiling heat transfer between the liquid and the tube. The result is that a significant portion of the coil surface purchased is doing little or no cooling.
- For the expansion valve to be able to control the liquid flow into the evaporator it is necessary for the plant to be operated with a fairly high head pressure. This is because the expansion valve is designed to control with a small variation in the applied pressure differential, i.e. the difference between the condensing pressure and the evaporating pressure. If this pressure differential were to be significantly lower the valve would probably starve the evaporator of refrigerant, and the evaporating pressure would be reduced with a consequent rise in running costs of the plant. During winter months with low ambient temperature it is necessary for the plant to be run with an artificially high condensing pressure, leading to increased costs.

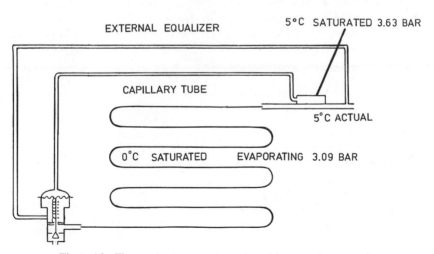

Figure 4.2 Thermoplastic expansion valve with external compression.

Other problems with these valves include:

- The setting-up of the superheat is difficult and invariably the valves operate with a higher than design superheat due to instability at lower superheat settings. This reduces the heat transfer for the evaporator, and depresses the evaporating temperature with a consequent rise in running costs
- The valves are unreliable and require regular maintenance or replacement

Even though the cost of the refrigeration plant associated with the direct expansion system is low, the increased lifetime operating cost of the plant will probably outweigh the savings in initial capital cost achieved by using the thermostatic expansion valves.

Gravity recirculation systems

This system is akin to the low-pressure system frequently used in liquid coolers and is probably more commonly referred to as a flooded system. The schematic layout for such a plant is shown in Figure 4.3. This shows a surge drum adjacent to the coil and installed in such a way that liquid can circulate freely from the surge drum and through the coil by gravity. The liquid is supplied from the high-pressure side of the system through a low-pressure float valve which maintains the level in the surge drum. Gaseous refrigerant is separated from the liquid in the surge drum, after it has passed through the coil, and is then drawn into the compressor suction.

This system generally results in circulation rates several times those required for complete evaporation. The result is that the evaporator surface is always completely wetted, leading to a high heat transfer coefficient. Because of this the entire evaporator coil surface is utilized to the full, making best use of the capital spent on the coil.

The expansion device used in ths system can be a simple float valve which

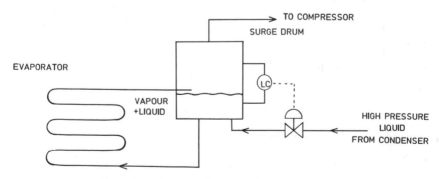

Figure 4.3 Gravity recirculation flooded evaporator.

controls the level of the liquid in the receiver. Being a fairly simple device it is robust and unlikely to go wrong. The additional benefit is that it can be operated with low differential pressures. This means that in the winter months with low ambient temperatures the plant can condense at as low a temperature as possible. The limitations will be the sizing of the expansion device and the size of the condenser and heat rejection equipment. It is quite possible for a cold store refrigeration plant to be operated with a condensing temperature of around 5°C during the winter months. The result is that the running costs are considerably less than the equivalent direct expansion system. A disadvantage is that the extra capital cost of the equipment associated with the surge drum will initially appear high. However, this will soon be recovered by the low running costs.

Where space near the evaporators is tight, or there is insufficient headroom, possibly in a cold store fitted into an existing building, it may not be possible to get the surge drum in the right place to operate. In this case a pumped circulation system could be used.

Installations utilizing multiple evaporators combined with gravity circulation systems require that each evaporator has its own surge drum.

A problem with the flooded system is the oil return from the surge drum to the compressor. Thus is considered in detail later.

Pumped circulation systems

The pumped circulation system operates in very much the same way as the gravity circulation system. Liquid is drawn from the bottom of the surge drum into a mechanical pump and pumped through the evaporator (Figure 4.4). The evaporated gas and overfed liquid is separated out in the surge drum where the gas is drawn into the compressor and the liquid stays in the receiver to be recirculated along with any fresh liquid admitted by the level control device.

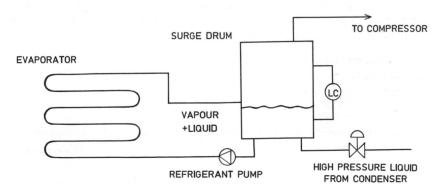

Figure 4.4 Pumped recirculation flooded evaporator.

Flash gas from the expansion valve (which is of no use in the cooling coil) is also separated and drawn off to the compressor.

The liquid can be fed into the evaporator from either the top or the bottom. The separation vessel or surge drum need not be at the same physical height as the evaporator and can be some distance away. The exact location determines the pump power required, which should be minimized to keep the running costs to a minimum; the shaft power is converted into heat immediately after the pump and uses up some of the refrigerating effect of the liquid being circulated.

The circulation rate is generally between two and six times that required for full evaporation (depending upon the refrigerant) and, as for the gravity circulation, the heat transfer coefficient is high leading to good use of the evaporator surface. The same comments about energy use for the gravity system are applicable to this system.

A disadvantage of this system over the gravity system is the reliability of the pump which is required to circulate the liquid through the plant. This is a minor problem, because modern pumps are normally very reliable. The benefits over the gravity system are that the pipe sizes required are smaller; friction losses in the gravity system will affect the circulation rate and consequently affect the plant's performance, so they have to be generously sized. With pumped circulation systems only one surge drum is required for many evaporators; this is a major advantage over the gravity circulation system, reducing the number of surge drums required and the complexity of the oil return system.

4.7   Capacity control of evaporators

As the duty required of the evaporator decreases it is necessary to reduce the amount of cooling that it can do. This is achieved by reducing the capacity of the compressor on single-evaporator systems. On multiple-evaporator systems this may not be possible as the evaporators will be operating at different loads and temperature requirements and as a result the capacity of each evaporator will have to be controlled separately. There are a number of ways to control the capacity of the evaporator:

- Cycling the plant or fans
- Back-pressure valves
- Variable speed fans
- Dampers

The simplest way to control an evaporator is to cycle the fan on and off. The temperature control in the cold space will be coarse, but in most cases adequate.

The back-pressure valve will impose a restriction at the suction line from the evaporator with the effect that a reduced amount of vapour will be drawn off from the evaporator and hence its duty will be reduced. The compressor will operate at a lower pressure than the evaporator and hence this method is not efficient in its use of energy.

By reducing the air flow through the evaporator (which can be done by reducing the fan speed or throttling the flow with dampers) the amount of heat required to be rejected by the evaporator will be reduced. This method is, however, limited in the modern cold store which requires the air to be thrown to the back of the store by the high velocity of the air leaving the coils.

In terms of energy consumption the above methods of capacity control can be ranked in decreasing energy efficiency as follows:

- Plant or fan cycling
- Air flow reduction
- Back-pressure valve

4.8 Oil return

Oil, if allowed to accumulate in the low-pressure side of the system, can play havoc with the controls and the performance of the evaporator. There are two distinct methods of treating the problems of oil return or separation for ammonia and CFCs. In spite of all attempts at reasonably efficient oil separation there is always a small amount of oil carryover, and the oil will find its way into the low side of the system and accumulate in the evaporator and surge drum if it is not continually removed from the low-pressure side of the plant. On low-temperature two-stage systems the best possible oil separator should be used to prevent the oil accumulating in the intermediate vessel and impairing its performance.

In general with direct expansion systems the design of the pipework ensures that the oil leaves the evaporator and returns up the pipework to the compressor.

With flooded systems the oil will accumulate in the surge drum and evaporator and cannot return with the suction gases. In small quantities oil is fully miscible with CFCs at high temperatures, but at low temperatures the oil will float on top of the liquid refrigerant as an oil-rich layer. At all temperatures oil separates out from ammonia, with the ammonia floating on top of the oil. At low temperatures the oil, if present in any concentration, will become waxy and interfere with the controls as well as affecting the heat transfer performance of the evaporator. It is therefore essential to remove the oil from the low-pressure side of the system.

With CFCs this is generally done using a rectification system in which a small portion of the liquid, oil-rich, mixture is drawn off from the surge drum and evaporated using a heat source (high-pressure liquid or electricity). The vaporized CFC and oil droplets are drawn into the compressor suction and the oil is thus returned from the low-pressure system.

For ammonia systems a separation space is provided at the bottom of the surge drum in which the oil can accumulate. From here the mixture of oil and ammonia is drawn off into an oil receiver located in a warm area in the plant room. The top of the receiver is connected to the separation space of the surge drum. The ammonia in the warm environment evaporates off, leaving oil behind. The oil can then be drained off at regular intervals.

Exact details of these oil recovery devices have not been described here, merely the principles behind the operation. This subject is extremely complex for the multistage, multicompressor units which require oil separation at every stage of the circuit.

4.9 Compressors

The compressor is one of the essential parts of a refrigerant system. There are two main categories of compressor: positive displacement compressors, which include reciprocating, rotary vane, and screw compressors; and the turbo-compressors, of which the centrifugal compressor is the only example. The operating temperatures and range of sizes is shown in Table 4.1. The purpose of the compressor is to draw refrigerant vapour from the evaporator, thus lowering the pressure and causing the liquid to boil, extracting heat from the cooling load in so doing at the desired temperature. The pressure must then be raised by the compressor to a level whereby it can be condensed by the available cooling medium.

4.10 Presentation of compressor performance data

The cooling duty and the power requirements of a compressor depend principally on the evaporating and condensing pressures. The performance of compressors is usually presented by the manufacturers as graphs, as shown in Figure 4.5, or as tables of duty and power over the range of evaporating and condensing pressures. These data are valid only for full-load operation with the stated suction superheat and liquid subcooling (the British Standard for the presentation of compressor performance data, BS 3122 part 2, specifies zero sub-cooling and 10°C suction superheat). Correction factors for differing conditions, part-load operation, and alternative speed operation are needed to obtain data corresponding to the actual operational conditions. Unfortunately these data, which are required by the British Standard, are not always readily available from the manufacturer, but it is essential that plant designers

Table 4.1 Typical compressor capacities.

Compressor type	Size range	Refrigerants
Reciprocating	0–1000 kW	R12, R22, R502, R717, R13, R13B1
Screw	50–2500 + kW	R12, R22, R717
Centrifugal	500–2500 + kW	R11, R12, R114
Rotary Vane	0–200 kW	R12, R22, R717

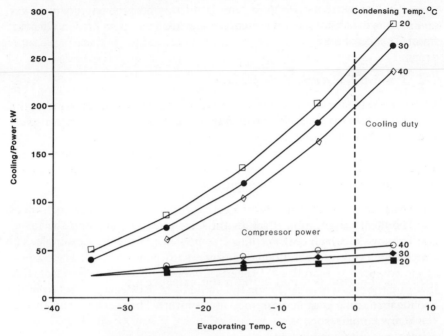

Figure 4.5 Typical compressor performance data.

and users obtain them. These data, provided to a common base, are particularly relevant where comparisons between compressors are required, for instance during the tender appraisal for an installation.

4.11 Compressor drives

The great majority of the prime movers involved with driving refrigeration plant are electric motors which have the distinct advantage that they are readily available, readily understood, and usually trouble-free.

There are three main configurations for the drive motors:

- Open
- Semi-hermetic
- Hermetic

An open compressor is one in which the compressor is totally free from the prime mover. A semi-hermetic compressor is one where the motor casing is directly coupled to the compressor casing, eliminating the rotary seal. Hermetic compressors are similar to semi-hermetic compressors in that the motor and compressor are mounted on the same shaft. The whole is then mounted in a welded steel case. As a result it is not normally possible to do any repairs to hermetic compressors and any failure usually means that the whole compressor has to be replaced. Most semi-hermetic and hermetic compressors have motors which are suction gas-cooled and are hence possibly less efficient, especially at part load.

4.12   Positive displacement compressors

A positive displacement machine is one where the increase in pressure of the refrigerant vapour is attained by reducing the volume of the compressor space by a fixed amount through work applied to the mechanism.

The performance of a machine is a measure of its ability to perform an assigned task. In the case of the compressor the performance is the result of a series of design compromises attempting to provide:

- The greatest trouble-free life expectancy
- The most refrigeration effect for the least power input
- The lowest capital cost
- A wide range of operating conditions

The types of compressor that fall into this category are:

- Reciprocating compressors
- Screw compressors
- Rotary vane compressors

4.13   Reciprocating compressors

These are the most commonly used type of refrigeration compressor and are available for a wide range of duties, as shown in Table 4.1.

System practices

The efficiency of the compressor is affected by the performance of the valves. The valve design is optimized for the refrigerant and the pressure ratio used. The compressor should not be operated with any other refrigerant or over a different pressure ratio before ensuring that suitable valves are installed. Operating a compressor with valves designed for low-temperature applications at high temperatures could result in a 5–10% reduction in cooling duty and a 20% reduction in compressor efficiency.

If a system using CFC refrigerant stands idle for any length of time the oil will become saturated with the refrigerant, which is to some extent soluble in the oil. When the compressor starts the pressure in the crankcase will drop, with the result that the refrigerant will boil out of the oil. This causes the refrigerant/oil mixture to foam, losing its lubricating properties and possibly allowing significant quantities of oil into the cylinders. To prevent this compressors used with CFC refrigerants are fitted with crankcase sump heaters which maintain the temperature of the oil when standing at between 50 and 60°C. Ammonia is much less soluble in oil than the CFCs and a crankcase heater is not required. No attempt should be made to economize by turning off the crankcase heater while the plant is idle.

Capacity reduction

The cooling capacity of a reciprocating compressor can be reduced by any of the following methods:

- Cylinder unloading
- Suction throttling
- Hot gas bypass
- Variable speed drives

The capacity of a reciprocating compressor can be changed by 'unloading' cylinders. This is done by holding open the suction valve so no compression is done by these cylinders (this unloading system is often used on large machines to enable them to be started off-load.) Using this method, the compressor capacity can be unloaded in fixed steps—typically 100%, 75%, 50%, and 25% of full load. This is a fairly efficient way of operating at reduced load; a compressor with half the cylinders unloaded will consume about 60% of the power of a fully loaded machine over the same pressure ratio (Figure 4.6).

Suction throttling reduces the suction pressure at the compressor manifold and as a result the compressor capacity is reduced (see Figure 4.6). The COP of

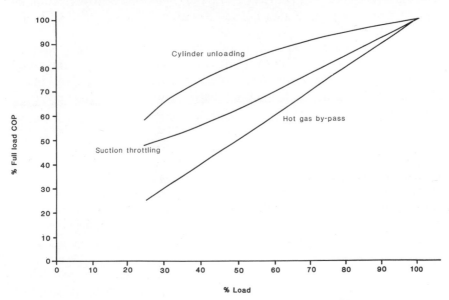

Figure 4.6 Effect of capacity reduction on system COP.

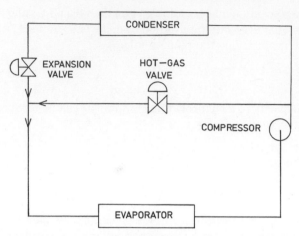

Figure 4.7 Hot gas by-pass.

the plant drops as a result of this suction throttling. This method can be used in conjunction with cylinder unloading but great care must be given to the interaction of two control systems. This control technique is extremely inefficient and should never be used.

Hot gas by-pass puts an artificial load on to the evaporator by by-passing some of the gas around the condenser (Figure 4.7). The result is that the compressor is still handling the same amount of gas but the cooling duty achieved is much reduced. The COP of a system operating with this method of capacity control is poor (Figure 4.6) and if at all possible this method should never be used.

A superior method of part-load operation is now available using variable speed drives which allow fully variable capacity control within the speed limitations of the compressor. Typically, turndown to 40% of full load is possible without cylinder unloading (dual-speed motors are a small step in this direction).

Maintenance problems affecting efficiency

To maintain the operation of a reciprocating compressor at its peak efficiency regular inspections of the cylinder heads are required to ensure that the valves are not damaged. The valves can fail due to liquid return to the compressor or through fatigue. Once failed they do not operate effectively and the capacity of the compressor is reduced with little reduction in power consumption. High discharge temperatures are a tell-tale sign of valve failure.

The wearing of cylinder liners or pistons has the same effect on the capacity and efficiency of the compressor. It is important, therefore, to ensure that regular inspections are carried out and necessary maintenance work performed to maintain the efficiency of the compressors.

4.14   Rotary vane compressors

Large rotary vane compressors are used in low-temperature applications as high-volume booster compressors. For reasonable efficiency they must be operated at or near their design condition. Typically the compressors are oil injected, which acts as both a lubricant and a coolant. These compressors are relatively compact and lightweight. The oil recovery and distribution system adds to the complexity of the plant.

Small rotary vane compressors are available which cover a wide range of duties and are used for small air-conditioning and refrigeration applications. They are of similar complexity and power consumption to other types of compressors. Their compactness and quietness play an over-riding factor in their choice for particular applications.

Unloading rotary vane compressors

The loading on some designs can be varied from 20% to 100% with a proportional reduction in the power consumption of the compressor. Unloading is achieved by porting slots in the end casings of the compressor. The slots are helical and cover the full length of the rotor, permitting complete venting of the rotor without damage to the blades.

4.15   Screw compressors

Screw compressors are available for duties ranging from several thousand to a few tens of kilowatts and can be used as low- or high-stage machines. The geometry of the machine determines the optimum pressure ratio and manufacturers usually produce a range of machines with different pressure ratios. It is important to note that the use of a machine with an incorrect built-in pressure ratio will result in significant loss of efficiency.

Nearly all screw compressors are oil flooded and, because large quantities of oil are carried over with the discharge gases, an oil separator is required. Because the oil and refrigerant are in intimate contact during the compression cycle, a significant proportion of the heat of compression is transmitted to the oil, and so an oil cooler is required to reject this heat. The oil can be cooled by either water or refrigerant. If the oil is refrigerant-cooled, this should be done with high-pressure liquid drained from the condenser; the vaporized refrigerant is returned to the condenser. If low-pressure liquid is used, then 10% of the capacity of the plant is used in cooling the oil with a consequent drop in overall system efficiency. Water-cooled oil coolers can utilize the condenser water or, alternatively, can be used for heat recovery; this is discussed later.

Capacity reduction

The capacity of screw compressors is variable from 100% to 10% of full load capacity on large machines by means of the sliding inlet valve. This reduces the compressor displacement by retarding the point at which compression starts and consequently reducing the volume of gas compressed. The part-load performance of the compressor is not particularly good—at 25% load the compressor takes about 40% of the full load power at constant evaporating and condensing pressure.

Single screw compressors

Although very different in geometry to the twin screw compressor, its operating aspects in terms of built-in volume ratio and part-load performance are similar.

Maintenance problems affecting efficiency

While the twin screw compressor has relatively few moving parts compared to the reciprocating compressor, there is one major area where loss of efficiency can and does occur. The rotor-to-case clearances are very tight, to ensure small slip losses. Any wear affects the efficiency and capacity of the plant dramatically with typically 25% loss in capacities recorded after six years' running.

Similar problems occur with the single screw compressor. Wear of the star wheels can lead to large losses of efficiency and capacity of the compressor.

It is important that the performance of the plant is carefully and regularly monitored to pick up any falling performance of the machine and to rectify the faults as soon as practical.

4.16 Centrifugal compressors

Centrifugal compressors are a member of the turbo-machinery family which include pumps and fans. They are widely used in packaged water-chilling plant because of their small size, low vibration, and relative simplicity. Centrifugal compressors have greater volumetric capacities, size-for-size, than do positive displacement devices. Their speeds are high for effective momentum exchange, 1800 to 90 000 rpm, but because the moving parts are rotating with no physical contact there is little vibration or wear.

Centrifugal compressors can be mounted with more than one impellor stage on one shaft. This can lead to compact compressors serving large duties.

Capacity reduction

The capacity of the compressors can be reduced by inlet guide vanes. Variable-geometry diffusers are sometimes used separately or in conjunction with guide vanes. Alternatively, variable-speed drives can be used to reduce the capacity of the compressor.

A major problem when reducing the capacity of centrifugal compressors is surging. This is the rapid flow backwards and forwards through the impellor of refrigerant gas, which limits the reduction in volumetric flow of the compressors. To overcome this, hot gas by-passing is employed to artificially increase the load on the evaporator and maintain the volume flow through the compressor. When a system is designed it is important that the surging is taken into account as prolonged operation with this condition can cause motors to burn our or physically damage the compressor.

4.17   Part-load operation and performance

So far in the sections on compressors the unloading mechanisms characteristic of each type of compressor have been discussed. A summary of the unloading techniques discussed so far is a follows:

- Reciprocating
  - Cylinder unloading
  - Variable speed
  - Suction throttling
  - Hot gas by-pass
- Screw compressors
  - Inlet slide valve
  - Variable speed
- Centrifugal compressors
  - Inlet guide vane
  - Variable geometry diffuser
  - Variable speed
  - Hot gas by-pass

Of all the types available, the variable-speed devices are by far the best way to reduce the capacity of the compressor, when these systems are correctly applied, with minimum loss in efficiency of the system.

In general, and depending on the method of control, when refrigerating systems are operated at less than full load the temperature differences in the evaporator and the condenser are reduced and the pressures in the heat exchangers rise and fall respectively. Both these effects raise the theoretical COP of the system. Against this the frictional losses in the compressor become a higher proportion of the total power and this will tend to reduce the COP. In addition to this the COP of screw and centrifugal compressors will be reduced due to the operation away from the full-load design point.

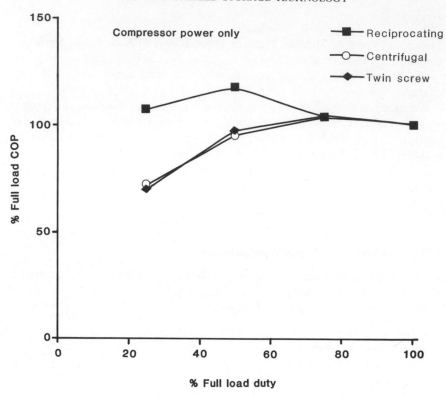

Figure 4.8 Part load compressor performance.

Part-load COP is therefore a balance between increased efficiency due to a smaller temperature lift and decreased efficiency due to increased losses. Figure 4.8 shows how the balance works out for typical reciprocating, centrifugal, and twin screw compressor systems. The poor part-load performance of the centrifugal and screw compressors can easily be seen. Figure 4.8 shows the performance of the compressors only, Figure 4.9 shows the effect of part-load operation on system performance which includes the power taken by the ancillaries, etc.

The compressor performance, and hence power cost, is only a part of the total ownership cost of a refrigeration system. The minimum cost can only be achieved by good system design and may turn out to use any of the types of compressors mentioned above.

4.18   Condensers

The purpose of a condenser is to reject the heat from the refrigerant system, which has been absorbed in the evaporator and input by the process of

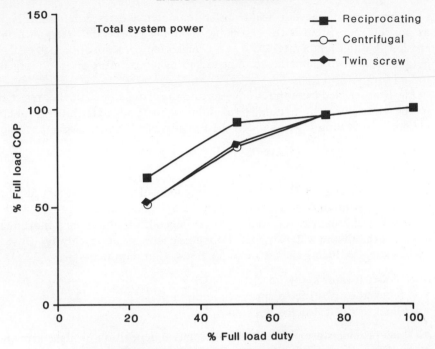

Figure 4.9 Part load system performance.

compression of the refrigerant, to the cooling medium. The gaseous refrigerant is thereby converted back to the liquid phase at the condenser pressure and is available for re-expansion into the evaporator.

There are three main types of condensers in widespread use:

- Water-cooled shell and tube
- Air cooled
- Evaporative cooled

The aim when designing and operating a refrigeration plant is to minimize the condensing pressure and thus use the minimum compressor power. With some types of systems this minimum condensing pressure is limited, due to the system design.

Water-cooled shell-and-tube condensers

Shell-and-tube condensers are used for all types of refrigerants, normally with the condensing refrigerant on the shell side and the cooling medium in the tubes. The tubes for condensing ammonia are normally bare tubes with no fins, but for the CFCs extended surface finned tubes are sometimes used to increase the heat transfer on the refrigerant side of the tubes.

To give a good tube-side heat transfer, the water velocity should be as high as possible, consistent with reasonable pumping power and freedom from erosion. A water temperature rise of 5°C and a temperature approach of 5°C between water exit temperature and condensing temperature are good targets to aim for.

The heat rejected from the condenser into the water can be further rejected to the atmosphere by a number of different methods. The most popular method is a cooling tower, though the use of well or river water is possible.

Cooling towers

The cooling effect is achieved by evaporating some of the water into the air; the latent heat required comes from the remaining water which is consequently cooled. The efficiency of the cooling tower depends on obtaining a thorough mixing of the air and water streams. Poor efficiencies are almost always caused by blockages reducing the air or water flows. Common faults are:

- Blocked water spray nozzles
- Blocked or obstructed spray elimination plates
- Blocked or obstructed packing

All these problems are normally due to mineral deposition or algae growth. Mineral deposition is usually controlled by water treatment and periodic or continuous blowdown which limits the concentration of minerals in the water. Algae growth is controlled by dosing the water with biocides.

Typically a combination of a shell-and-tube condenser and cooling tower should result in condensing temperatures which are 15°C above ambient wet-bulb temperature in the summer and 18°C above wet-bulb temperature in winter.

Cooling by well or river water

Well water remains at a reasonably constant temperature throughout the year, typically at about 10°C. Extracting this water and using it in the condenser can lead to reduced running costs for the plant due to the consistently lower condensing temperature achievable. River water in some locations can also remain at a reasonably constant temperature through the year, though in many locations the temperature will vary by 10°C or more. The river water temperature is, however, lower than that obtainable from the cooling tower. This is therefore a good source for a cooling medium into which the heat from refrigerating plants can be rejected. Other possibilities are lakes, docks, or the sea.

Where condenser fouling could be a problem there are several heat exchanger designs which can be used to minimize the effects.

Air-cooled condensers

The air-cooled condenser rejects the heat directly from the refrigerant to the air. The refrigerant condenses inside the tubes over which air passes. Because the air-to-tube heat transfer is poor, the tube surface is invariably extended using metal fins. Very often air-cooled condensers are situated remotely from the plant room; this involves lengthy pipe-runs, with consequent higher probability of losing refrigerant through leaks.

Without condensing temperature limitations, a temperature difference of 14°C between the condensing temperature and air inlet temperature is normally economically achievable.

Evaporative condensers

The evaporative condenser combines a cooling tower and refrigerant condenser in one piece of equipment (Figure 4.10). A fan forces air through the condensing coils and the falling water. Water is continuously pumped from the pan and sprayed over the coil. A small proportion of the water is evaporated, removing the heat from the refrigerant and condensing it inside the coil. The advantage of the evaporative condenser over the cooling tower is that large volumes of treated water are not pumped continuously around the cooling circuit, reducing considerably the pumping and water treatment costs.

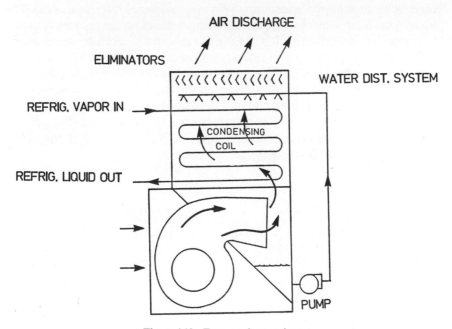

Figure 4.10 Evaporative condenser.

The performance is similar to the shell-and-tube condenser/cooling tower combinations. A condensing temperature 15°C above the ambient wet-bulb temperature in summer and 18°C above wet-bulb temperature in winter is achievable. The operating costs will be less through the reduced pumping power required, and the initial capital cost will be lower than the shell-and-tube condenser and cooling tower combination.

4.19   Head pressure control

If any of the condensers or combinations of equipment were allowed to run uncontrolled in the winter conditions, very low condensing pressure ('head pressure') could be encountered. This is particularly important for systems using thermostatic expansion valves which rely on a relatively constant pressure differential to be able to control the refrigerant flow. For thermostatic expansion valves it is necessary to operate the plant with summer condensing pressures during the winter months. This is the penalty for purchasing the 'cheap' expansion device. An example of the increased COP of plants operating at a high head pressure compared to a low head pressure is shown in Table 4.2.

For systems which do not require this high head pressure to operate it is possible to make full use of the low ambient temperatures during the winter months by allowing the condensing pressure to drop as low as practical. The limitations on this are the pressure drop required across the expansion device to get the liquid flow through it. At the design stage this can be overcome by sizing the valves generously or using two or more small valves in parallel. On existing plant a small boost pump in the liquid line to assist with the refrigerant flow through the expansion device is an alternative. In these instances cooling tower temperature control would still be required, to prevent the minimum pressure from being attained or to prevent the tower from freezing up.

Head pressure control for these systems can be achieved in a number of ways, depending on the type of heat rejection equipment.

Table 4.2   Head pressure control: COP at different condensing temperatures.

Condensing temperature (°C)	Evaporating temperature (°C)	COP
35	−25	2.1
5	−25	5.2

Shell-and-tube condensers and cooling towers.

There are three main methods of head pressure control on these plants:

- *Cooling tower fan cycling* operates by varying the speed of the fan or starting or stopping it in accordance with the signal from a thermostat immersed in the water leaving the tower. This method of control is quite coarse, leading to quite large temperature swings. An alternative method used on some towers is to throttle the air flow using dampers, which can give better temperature control.
- *Cooling tower water bypass* operates by by-passing some of the warm water flowing to the tower and mixing it with the water from the tower. This is often used in conjunction with fan cycling to obtain better condenser water temperature control. Thermostats to control the fan operation will still be required to prevent overcooling and possible freezing of the water.
- *Condenser water flow control* operates by throttling the water flow on the outlet of the condenser. The flow control valve can be controlled from the refrigerant pressure using a pilot-operated valve.

Air-cooled condensers

Head pressure control is achieved by cycling fans on and off on multiple fan units, controlling the air flow with dampers with a smaller numbers of fans, or 'flooding' some of the circuits by holding back the liquid flow.

Evaporative condensers

Head pressure control is achieved by:

- Cycling fans on and off
- Variable-speed fans
- Damper control

In addition, reducing the water flow over the tubes will reduce the cooling capacity of the condenser. This is used particularly when freezing of the water is possible, in which case the water circulation pump is turned off.

4.20 Factors affecting condenser efficiency

Surface fouling

Shell-and-tube condensers Build-up of mineral deposits or scale on the water-side of the condenser inhibits the transfer of heat between the refrigerant and the water. Proper treatment of the cooling water and regular cleaning will reduce the effect of fouling. If fouling is known to be a problem then the design of the condenser should be able to accommodate fouling and facilitate regular cleaning.

Air-cooled condensers   Fouling on these condensers is invariably due to leaves or other matter becoming trapped on the fins on the condenser. If it is not practical to eliminate the source of the fouling then regular cleaning with a brush will ensure that the running costs are minimized.

Evaporative condensers   The main cause of fouling is through the build-up of mineral deposits on the coil surface. Regular blowdown and adequate water treatment should eliminate the need to clean the surfaces. If deposits are allowed to accumulate, the surface can be very difficult to clean.

Accumulation of non-condensible gases

Air and other non-condensible gases can get into the refrigeration system in several ways. The most common reason is insufficient evacuation of vessels and pipework prior to initial charging or after maintenance. In systems which operate at sub-atmospheric pressures in the evaporator (for example R22 below $-40°C$), leaks will be inward rather than outward and large quantities of air can get in over a period of time. In any system there will be a breakdown of the refrigerant and this too results in a build-up of non-condensible gases.

The build-up of air or other non-condensibles results in high condensing pressures and high apparent liquid sub-cooling in the condenser. Their effect on operating efficiency can best be seen by an example.

If the condenser contains 15% air and 85% ammonia and the absolute pressure is 15 bar, the partial pressure of ammonia is 12.75 bar. Therefore, because of the presence of non-condensibles, the discharge pressure is 2.25 bar greater than it would be if there was only ammonia present. The operating cost in this example will increase by 12% if the evaporating temperature is $-10°C$.

Testing the presence of non-condensibles is relatively straightforward in principle. The general method for testing for non-condensible gases is as follows. The condenser has to be isolated on the refrigerant side and a pressure gauge, which should be calibrated, installed. The cooling medium, be it air in an air-cooled condenser or an evaporative condenser (the water should be turned off), or water in a shell-and-tube condenser, should be left running. If there are no non-condensibles present then the gauge pressure will be the saturation temperature at the air/water temperature. With the non-condensibles present the gauge pressure will be higher than the saturation temperature of the air/water mixture.

The procedure for purging non-condensible gas from the condenser should be described in the plant maintenance manual.

4.21 Location of plant room

A number of factors should be taken into to account when locating the plant room:

- Distance to evaporators (pipe lengths)
- Distance to condensers (pipe lengths)
- Ease of access for maintenance
- Safety with refrigerants

The pipe runs, be they refrigerant of fluid (water/brine), should be minimized to reduce the capital cost and the running costs. An optimum solution for this can be found in terms of balancing the running cost with the capital costs.

A plant room tucked away in an inaccessible place is ideal aesthetically but can prove very costly if items of equipment need to be regularly removed. It is not advisable to build the plant room up around the equipment, as it will then be difficult to replace the equipment even with similar sized equipment should this be needed.

If a plant room has to be in a basement then full precautions for the build up of refrigerant gas should be taken. The necessary safety requirements of BS 4434 should be adhered to at all times.

4.22 Heat recovery from refrigeration plant

Refrigeration plants are potentially a source of heat for use on the site. Generally the heat rejected to the atmosphere through the cooling water is at a temperature too low to be of any real use; for the efficient operation of the refrigeration plant the temperature lift from the evaporator should be as low as possible and as a result the temperature at which heat is rejected should be as low as possible. Therefore to recover heat is at the expense of the refrigeration cycle efficiency if the condensing temperature must be raised. If the extra running cost of the plant operating at the elevated condensing pressures is offset by the savings in capital and running costs of the heat user, then heat recovery could be cost-effective. Heat recovery is not a 'bolt-on' extra for a refrigeration plant, and should be a fully integrated part of the design of the system.

There are several methods by which the heat recovered can be at a useful temperature. This topic is considered later. However, it is important to consider first what the heat is to be used for.

The following are potential heat sinks:

- Space heating
- Domestic hot water or washdown water
- Frost heave prevention

Conventionally space heating for offices, etc. is provided by a gas- or oil-fired boiler for heating low-temperature hot water with a temperature programme of 80°C flow and 70°C return. These are the temperatures used for conventionally sized heating equipment. The system can be operated at a lower temperature by oversizing the heating equipment. Space heating is required only during the winter months.

Domestic hot water and washdown water is typically required at 60°C and can be supplied by a calorifier heated by fossil fuel or by electricity. The quantities are small but they are year-round uses.

An important part of the construction of cold stores in particular is the prevention of frost heave. This is achieved by installation of a heated floor which is heated either with electricity or with warm liquid (often glycol) circulated through pipes in the floor. This requirement is year-round and the temperature of this liquid does not have to be very high.

The heat to satisfy these sinks can be recovered from the following:

- Discharge gas de-superheaters
- Screw compressor oil coolers
- Condensers

These are considered in turn.

Discharge gas de-superheaters

Because of the properties of ammonia, it is more suited to this method of heat recovery than the CFCs. The discharge temperatures of ammonia, R22, and R502 are shown in Table 4.3 operating over the same cycle conditions. Typically, if the ammonia gas is cooled to 45°C, water can be heated up from 15°C to 65°C; under this condition 20.4 kW of heat are available for every 100 kW of cooling duty. Heat recovery from the CFC cycles is possible but the temperature of the water from the de-superheater will be much lower or the water flow must be lower to obtain the higher temperature. This heat recovery method will ideally be suited to small heat sinks such as domestic hot-water heating. Any supplement can be provided by an electric heater.

In practical installations, the de-superheater should always be located above the condenser so that if any liquid refrigerant does form then it can drain freely into the condenser.

Screw compressor oil coolers

In single and twin screw oil-flooded compressors a considerable proportion of the motor power is dissipated as heat in the oil. This percentage is of the order of 38% for R22 and 60% for ammonia systems. The compressors are usually designed for an oil inlet temperature of 40°C and oil exit temperature of between 60 and 80°C, though the exact temperature depends on the pressure

Table 4.3 Heat available from a de-superheater.

Evaporating temperature (°C)	Condensing temperature (°C)	Refrigerant	Cooling duty (kW)	Typical compressor power (kW)	Compressor discharge temperature (°C)	Sensible heat in super-heated gas (kW)	Heat of condensation (kW)
−15	30	R22	100	30.9	72.2	21.7	109.2
−15	30	R502	100	32.6	55.4	20.2	112.5
−15	30	R717 (Ammonia)	100	27.2	127.1	24.1	103.1

Table 4.4 The costs and benefits of condenser heat recovery.

Case	Condensing temperature (°C)	Compressor power (kW)	Heat rejected (kW)	Cooling water (°C) In	Cooling water (°C) Out	Cooling water flow (kg/s)	Electrical cost (£/h)	Value of recovered heat (£/h)	Net saving (£/h)	Relative compressor size for 100 kW cooling
Standard chiller	30	20	130	15	25	2.87	0.9	0	1	1
High condensing	45	30	130	15	40	1.24	1.35	1.87	1.42	1.13
High condensing with de-superheater	45	30	130	15	45.8	1.01	1.35	1.87	1.42	1.13
High condensing with de-superheater and sub-cooler	45	24.8	124.8	15	44.9	1.00	1.12	1.80	1.58	0.93

ratio. This source of heat is probably ideal for heating domestic hot water, or for frost heave prevention. The temperature is insufficient to provide hot water for space heating in conventional systems.

Condenser heat recovery

By far the greatest source of heat for heat recovery from refrigeration plant is the condenser. Typically a plant may operate with a condensing temperature of 30°C. In the condenser the closest the water can get to 30°C would be within about 5°C; 25°C water is not much use for anything except pre-heating boiler feed water.

By increasing the condensing temperature to 45°C the temperature of the water can be raised to 40°C, which is more useful. However, the efficiency of the refrigeration cycle will now be affected because the pressure ratio across the compressor has increased and the refrigerating effect of the condensed liquid is reduced due to the increase in the flash vapour produced across the expansion valve; the increased pressure ratio required to obtain these temperatures may not be possible due to the limitations of the compressor. The refrigerating capacity is typically reduced by 15.5% and the power increased by 27%.

The economics of increasing the condensing temperature are shown in Table 4.4 which assumes a fixed cooling duty of 100 kW. Case 1 is the base case which assumes no use can be made of the heat from the condenser and electricity costs 4.5p/kWh. Case 2 assumes that the condenser cooling water is heated up from 15°C to 40°C and that this heating is used to replace heat from gas at a cost of 33p/therm (3.13£/GJ) and a thermal efficiency of 78%.

Higher-temperature water can be obtained by using a separate heat exchanger to de-superheat the gas from the compressor discharge before it enters the main condenser. Water is heated up to within 5°C of the condensing temperature in the condenser before being fed into the de-superheater where the temperature is further raised. The net heat recovered is the same but the temperature is higher and therefore of more use. This is shown in Case 3.

The savings can be further increased by subcooling the liquid from the condenser in a water cooled heat exchanger fed with 15°C water. The subcooled liquid obtainable should be about 20°C. This will increase the refrigerating effect without increasing the compressor power. Case 4 shows this as a reduced compressor power for a constant cooling duty.

It is vital when designing a refrigeration system incorporating heat recovery to take into account the power taken by ancillary loads; for simplicity, this has not been done in Table 4.30. In addition it is important to consider the heat rejection from the condenser when recovered heat is not required. Raising the condensing temperature will greatly increase the required compression ratio, and care must be taken to ensure that the plant is operated within the manufacturing limits at all times.

4.23   Heat pumps

So far the discussion has been restricted to the recovery of essentially low-grade heat up to a temperature of possibly 50°C. This temperature is not very useful for space heating and will provide domestic water at a temperature below that required. The temperature of the recovered heat can be boosted by using a heat pump. In the previous examples the limitations on the temperature of the water are the maximum operating pressure and pressure ratio of the compressor. By using a separate compressor for elevating the condensing temperature and the temperature at which heat can be recovered more useful temperatures can be obtained.

The heat pump would operate with the evaporator as the condenser of the refrigerating plant. The heat pump would probably be evaporating at around 25°C and can then condense at an elevated temperature of say 60°C or higher. To be able to operate at these temperatures, a refrigerant which is not at a very high pressure and can be used with conventional equipment would have to be selected. This type of heat pump is known as a cascaded heat pump and can utilize different refrigerants in the second or third stages to those used in the other stages of the plant.

The design of the heat pump is critical for cost effectiveness. In general terms a heat pump is only cost effective if it has a high utilization throughout the year. On this basis it would probably be used for domestic hot-water production or space heating, if the alternative heat source is expensive to bring on site. The heat pump need not be sized to take all the heat rejected by the cooling plant, and in this way the size and cost of the plant can be optimized to suit the site.

4.24   Consideration of cooling loads

So far no mention has been made of calculating or reducing the cooling loads on a cold store. In this section the methods of estimating the required refrigeration plant capacity will be discussed, as well as additional factors which can be used to reduce the cooling load for the store.

When estimating the loads on a store the following factors have to be taken into account:

- Transmission heat gains
- Heat gains from air infiltration
- Heat from fans, refrigerant pumps (primary or secondary refrigerant), lighting, human occupation
- Heat from coil defrosting equipment
- Heat released by goods in storage
- Heat removed from goods in cooling down to storage temperature
- Heat to be removed in freezing unfrozen goods
- Blast freezing or process freezing
- Others loads not detailed

This is quite a long list: if all these possibilities had to be considered the refrigeration plant would have to be very large indeed and would be well over capacity for normal running. The first five items of the list are unavoidable loads on the cold store, and reducing them is the aim of many cold store operators. These will be considered first.

Transmission gains depend on the thermal conductivity of the material of construction of the cold store and the difference between the internal and external temperatures. The latter is beyond the control of the designer but the former can be optimized between capital and running cost of the store. Transmission gains can be calculated with good accuracy.

Air infiltration is extremely difficult to eliminate totally and, depending upon the use of the store, can represent 50% or more of the cooling load (a long-term storage warehouse will have a lower usage of the doors, reducing the infiltration load). Calculation of the infiltration rate of warm air through open doors is nowadays done using sophisticated computer programs. This removes some of the black art of estimating infiltration loads. These loads can be reduced by using quick-shutting doors, air locks, plastic curtains, or cold air blasts at the doors but none of these can eliminate the infiltration gains totally. An added problem with the infiltration of warm air into the store is the moisture that accompanies the air. This puts higher loads on the defrosting of the coils and consequently the running costs.

Heat from pumps can be accurately calculated once the pipe runs and sizes are known. This heat gain can be reduced by minimizing pipe lengths and maximizing the size. This requires careful optimization.

Fan power can be reduced by using efficient fans and optimizing the pressure drop through the heat exchangers by ensuring sufficient surface area exists to achieve the heat transfer without unnecessary pressure drop caused by high velocities.

The minimum permissible *lighting* levels, according to the codes of practice, determine the number of lights that are required in a store. Selection of the most energy-efficient lights can reduce the heat load from this source. Traditionally, fluorescent lights suitable for low temperature have been used in cold stores, though nowadays high-pressure sodium lights may be used. The ultimate aim is to reduce the losses from the control gear and still obtain the maximum light output. The most efficient way to operate a cold store is to turn the lights off when they are not needed!

The heat released by the *coil defrosting equipment* depends on the types of defrost chosen. Heat from the coils will pass into the cold space during defrost but this can be minimized by shutting down the fans during the defrost cycle and not starting them up until the coil has reached operating temperature again. Calculation of the heat gain depends on the frequency of the cycle and the method of defrosting, and is relatively straightforward.

The heat gain *human occupation* can be calculated readily, but due to the nature of the operation of a cold store cannot be reduced unless the store is to become fully automated.

Information of the heat release from goods in storage is readily available in ASHRAE and therefore the product storage load can be reasonably accurately calculated.

Due to the delay between unloading refrigerated vehicles and storing the goods in the cold store, it is inevitable that the goods will gain heat. This is detrimental to the product and also to the operation of the store. Goods for cold storage should be received at storage temperature and any transfer operations should take place in cold air locks. If warm products do enter the store then the bulk temperature will rise and affect all the products while the plant tries to cool down the store to its operating temperature. If it is known that this will happen and is unavoidable, the plant should be sized to cope with it.

Blast freezing or process cooling can require quite substantially sized plants. As a result it is not good practice to install a plant to serve both a cold store and a blast freezer unless the plant can be operated efficiently with a good level of turn-down. This would typically require a plant to have large number of compressors on the same refrigerant circuits.

4.25   Plant monitoring and control

Once a refrigeration system has been purchased, installed, and commissioned, the operating efficiency and overall running costs will be largely determined by the effectiveness of the day-to-day monitoring. Even when an outside maintenance contractor is used, the plant operators on the site are the only people who are able to detect faults at the early stages when they can significantly increase the energy consumption without preventing the system meeting the required cooling duty.

To monitor the performance of a refrigeration plant effectively it is necessary to have a minimum of plant instrumentation, which should be specified at the purchasing stage.

4.26   Instrumentation

At the very least the plant should be supplied with pressure gauges either side of the compressors, compressor ammeters, and kilowatt-hour meters; most systems to comply with BS 4434 have to be supplied with pressure gauges or gauge connections. This level of instrumentation is insufficient for performance evaluation or thorough fault diagnosis. In general terms, to be able to carry out any detailed performance testing, all pumps, secondary or primary refrigerant, evaporators, condensers, and compressors as well as any intermediate pressure vessels, should be fitted with pressure gauges or at least have a gauge tapping on both the secondary and primary refrigerant sides where relevant.

Accurate temperature measurement is a great aid to the efficient operation of refrigeration plant. It is, however, the easiest measurement to get wrong, and incorrect measurement is worse than no measurement at all. Two things have to be got right for success; the instrument must have sufficient inherent accuracy and the temperature 'felt' by the sensor must be the same as the fluid temperature. The accuracy of temperature measurement required in a refrigeration system is higher than that required in most other applications, and in general should be better than $\pm 0.1°C$. In general, the sensor must be located in a representative part of the fluid flow. This will normally require a thermometer pocket extending at least one third of the way into the pipe. The pocket should be filled with oil to give good heat transfer to the sensor.

Thermocouples and resistance thermometers, when coupled with suitable indicating electronics, are the most suitable for general temperature measurement. Platinum resistance thermometers are more accurate than thermocouples. A system incorporating either of these measurement systems permanently connected to the system and indicating on a central station is a good method of recording the temperatures around the plant.

4.27   Plant log sheets

The plant log sheets provide an effective way of gathering data on plant performance but they are only of any real use if:

- The data recorded are accurate
- The information is intelligently analysed
- Any problems identified are followed up

There is no such thing as a standard log sheet—each plant should have a log sheet made up to reflect its actual needs. On complex plants, several sheets may be needed with each sheet covering a separate part of the system.

The log sheets can only provide a rough guide to how the plant is operating. The ideal operating parameters of a refrigeration system vary with the evaporator inlet temperature, system load, and ambient temperature. The only way to satisfactorily determine the operating efficiency of a refrigeration plant is to perform a full heat balance and then compare the actual plant performance with the design performance.

4.28   On-line monitoring and plant diagnosis

The trend in recent years has been towards microelectronic monitoring and data logging systems. While this is a convenient method of recording data, it does mean that there will be fewer visits to the site to identify warning signs such as unusual noises and oil leaks. The calculations necessary to perform the heat balances mentioned above can be readily done by the system monitoring the plant. This will enable the analysis of the plant to be carried out more readily.

By regular monitoring of the performance of the plant, faults can be identified long before they can cause a problem. Fault finding on the refrigeration plant can be extended to be performed by microprocessor-based systems used to monitor the plant. The faults would be determined from a series of logical statements and the plant operators would be informed of any problems with the plant as and when they were identified by the microprocessor and before the users of the 'cold' realize that a plant is failing.

5 Store insulation

B.A. RUSSELL

5.1 Historical background

The need for insulation

During the early part of the twentieth century, the need for insulation of storage areas become apparent following experiments in marine transport. Before this, ship cargo areas were loaded with ice to keep products fresh during transit. Soon refrigeration systems and insulated enclosures were established to ensure that fresh produce would arrive at its destination in good condition.

Various forms of insulation materials were used; the earliest were wood chippings compacted between the ribs on the ships' sides and bulkheads and lined with either timber boards or metal cladding. It was soon realized that this form of insulation absorbed moisture to such an extent that eventually it became almost solid ice. Although the ships' sides and metal bulkheads protected the insulation from the ingress of moisture, when the refrigeration plant was turned off all surplus water was then absorbed into the insulation, and subsequently formed an ice build-up once the store was again reduced to temperature.

Earlier land cold storage relied primarily on a concrete structure, and, in some cases, when it became obvious that moisture penetration through the concrete was also causing an ice build-up, cavity walls were introduced. Consideration was not given at this time to an effective vapour seal. Cork slab was the most effective insulant, but eventually the insulated structure became ineffective due once again to moisture penetration.

In situ cold storage

Before the introduction of cork slab, there were few materials which could even be considered seriously in connection with cold store construction. For many years cork slab was the only acceptable material, having been developed in Spain and Portugal for insulation to ships' holds and cargo areas where more sophisticated refrigeration was being developed. Cold storage manufacturers soon realized that this material, proven in marine applications, would be ideal for the development of *in situ* cold storage units that were to form the basis of construction for industrial insulated enclosures well into the mid-twentieth century.

Once it was realized that cork could be processed into easy-to-handle slabs, and packed in sizeable containers for export, the cork-growing industry of Spain and Portugal expanded rapidly, supplying the cold store industry worldwide. The UK became one of the world's biggest importers of cork slab, and cold store manufacturers soon adapted this material for various uses in the design of *in situ* cold store construction.

Early design and techniques

These *in situ*, or 'built-in' stores as they were then known, were the most common construction forms utilizing cork slab. They consisted of brick or concrete structures built up from solid foundations and having a concrete sub-floor and concrete roof. In many cases, self-supporting ceilings were constructed (see Figure 5.1).

Walls and concrete ceilings were rendered with half-inch sand and cement to leave a smooth surface. When this rendering was dry, walls, ceiling, and floor would be treated with a vapour seal solution (vapour barrier). The essential need for a vapour protection was recognized after the dismantling of earlier stores, where the insulation was found to have absorbed moisture through the concrete or brick walls. This had turned to ice once the refrigeration plant had been commissioned.

The vapour protection seal consisted of a bituminous rubber solution applied to the prepared surface in two separate coats, by brush or trowel. When this had 'cured' the cork slab was applied to walls and ceiling with hot pitch and supported with timber grounds mechanically fixed to the walls. A second layer of cork was then applied with all joints staggered and sealed, ensuring no through joints. Additional mechanical fixings were made through this second layer of cork into the first layer of timber grounds (Figure 5.1).

The floor insulation was laid in a similar manner, but this would normally be left until after the wall and ceiling finish was complete. This avoided damaging the cork floor by scaffolding towers, etc. The most common finish for the walls and ceiling was plaster (subsequently decorated), or a polar white cement that needed no decorative finish. This would be trowelled onto an expanded metal, fixed to the cork and timber grounds by means of staples.

Other applications included glazed asbestos and galvanized or zinc-coated steel sheeting fixed to additional timbers embedded in the insulation. These materials became unpopular due to the method of fixings and the numerous cover strips, etc., required to achieve an acceptable apperarance. Most cold store users favoured the cement finish for its clean, smooth surface which could be easily redecorated every few years.

Normally the wall finish would terminate at a pre-determined height above

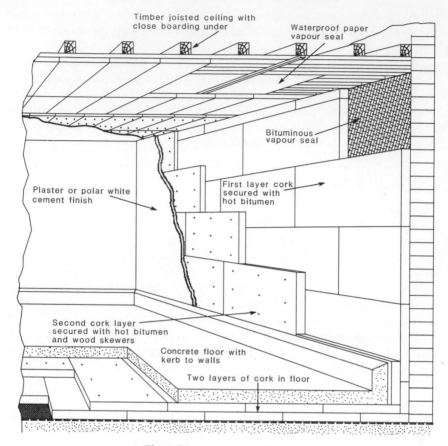

Figure 5.1 *In situ* or built-in store.

the planned floor level, to allow for installation of a cove or concrete kerb.

Depending on anticipated loading, a suitable thickness of solid granolithic internal flooring would be laid over the cork insulation, and metal filings and granite dust would be incorporated into the top half inch, trowelled smooth to provide a hard-wearing surface.

Conventional systems

During the late 1930s, developments in the frozen food industry created an urgent requirement for low-temperature cold stores. Cork continued in use primarily for insulating walls, ceilings, and floors of *in situ* stores, but a considerable increase in thickness was necessary to accommodate the lower operating temperatures required for the storage of frozen foods.

Prefabricated units were also constructed by the conventional system. These consisted of timber frames secured together in modules of 1200 mm width, covered externally with metal sheets or timber boarding. Cork insulation was fitted between the timber frames, and a bituminous paper or solution was applied between the insulation and the external cladding to form a vapour seal. The internal finish would normally be glazed asbestos sheeting or steel cladding. Because of the difficulty of transporting these very heavy units to site (the dense cork slab alone contributed greatly to the weight), such prefabricated stores were rarely more than 6 m high.

These early systems of cold store construction, generally accepted by the refrigeration industry, formed the basis for the principles of cold store design well into the 1960s and later, following the advent of panel construction, became known as 'timber and concrete stores'.

5.2 Modern developments

The end of the 1950s and early 1960s saw the development of two new products—expanded polystyrene (EPS) and polyurethane foam (PU).

Polystyrene was developed primarily as a low-cost, low-density material with thermal conductivity of $0.034 \, W/m^2/°C$. This proved to be a natural replacement for cork board, as it was produced in board form with the same dimensions as cork. The conductivity of cork slab insulation (density 8 to 9 lb per cubic foot), as used in the construction of cold storage, was rated 0.26 to $0.28 \, BTU/hr/ft^2/°F$. It was doubtful whether these values were achieved in practice, due to the inclusion in the insulation of timber beams, supports, and timber door frames, etc. With lower density polystyrene weighing approximately eight times less than cork, the steel supports necessary to carry a ceiling structure could be greatly reduced, making considerable cost savings.

As with all new materials, it was some time before polystyrene achieved acceptance as a substitute for cork board; subsequently, as manufacturers of cold storage became more innovative and panel production was developed, expanded polystyrene became the alternative to cork in the design of modern cold store panel production.

Polyurethane foam (PU), initially used for insulation of domestic refrigerators and smaller commercial rooms, either in board form or foamed *in situ*, was later developed through different applications for use in industrial cold storage panels.

One method of production was to form the polyurethane into blocks or buns from which different thickness slabs could be cut. These would then be used in composite panels. Another method was to inject the foam directly into a jig that contained the inner and outer faces of steel to form an *in situ* foam panel. Once polyurethane was accepted in the cold store industry as a good substitute for either cork or polystyrene, considerable advances were made in new techniques and further applications using this material.

5.3   Insulation materials

Expanded polystyrene

Expanded polystyrene (EPS) is one of the most efficient rigid insulation materials available today and is widely and successfully used throughout the cold store industry.

Derived from crude oil, by the combination of benzene and ethylene, which produces styrene monomer, EPS bead is created by the addition of catalysts and an expanding agent, pentane. In the bead form it is a sugar-like material. During the process known as pre-foaming, the raw material expands rapidly forming thousands of tiny cells within each bead; these hold the air captive and produce the EPS.

After conditioning, the pre-foamed bead is moulded to produce blocks up to $7500 \times 1350 \times 650$ mm in size. From these blocks are cut sheets and slabs in any required thicknesses to be used in the composite and continuous laminate processes of panel production.

EPS is manufactured in the UK to BS 3837: 1987. This defines the minimum requirements for such aspects of its performance as compressive and cross-breaking strength, thermal conductivity, water vapour permeability, and flame retardancy (see section 5.15).

Typical performance figures for SDFRA and HDFRA material used in the manufacture of cold store panels conform to the minimum requirements laid down in BS 3837. Extra heavy density materials (EHDFRA) are also used extensively in the industry where heavy loads are required. The letters FRA after the density denote flame-retardant additive or self-extinguishing grade materials).

Extruded polystyrene

Extruded polystyrene is basically manufactured from the same raw material as EPS, with the exception that extruded polystyrene for use in panel production is a foam insulation board without a skin. Other forms of extruded polystyrene are available, incorporating a skin, such as the heavier density used for floor insulation purposes. It is manufactured by a continuous extruding process whicn gives a rigid closed cell structure with unique properties.

It is an ideal material for the use of panel production in the cold store industry because of its high resistance to water absorption and its superior mechanical properties. The high resistance to water absorption and water vapour difffusion results from its closed cell structure, and the inherent resistance of the base polymer to water. The high resistance to water absorption enables the material to maintain a low thermal conductivity.

The high tensile strength of extruded polystyrene makes possible a good bond between the foam and the facing materials, and the high shear strength reduces the risk of failure in the panel core material. Because of its high compressive strength, extruded polystyrene, when bonded to a steel face, reduces the possibility of impact damage and is thus an excellent material for use in composite and continuous laminated panels.

Polyurethane and polyisocyanurate

Rigid polyurethane (PUR) foams are highly cross-linked polymers with closed cell structures which bubble within the material, with unbroken walls, so that gas movement is retarded. The chlorofluoromethane gas is contained within the walls and, as these substances have a much lower thermal conductivity than air, such closed cell forms have significantly lower thermal conductivity than any open cell foam.

However, to retain this low thermal conductivity the gas must not leak away; consequently, rigid foam insulation must have at least 90% closed cells and a density above $30 \, kg/m^3$. Rigid foams are made by the combination of a polyol and a liquid blowing agent, plus a catalyst, plus a polyisocyanurate (PIR). 90% of polyols used are poly-ethers with terminal hydroxyl groups, and the isocyanate used is di-isocyanato-diphenylmethane.

Polyisocyanurate foams are particularly important because of their resistance to high temperature and their relatively low combustibility. In the manufacture of these foams, the polyisocyanurate is polymerized to produce an isocyanurate ring structure which is thermally stable. All established polyisocyanurate foam systems are actually polyurethane-modified polyisocyanurates.

The most economical method of making large quantities of slab stock foam is on on a continuous manufacturing basis. This tends to give a product of high quality as it is easier to control the cell size and cell structure uniformity. The foam reaction mixture is dispensed continually into a trough formed by paper or a polyurethane film in a moving conveyor belt. The trough is designed to accommodate the foam pressure on the side walls, pressure that occurs just after the foam has risen completely. A flat-top block is obtained by the use of a top converter, or by assisting the rise of the foam by other processes. Blocks of any length and thickness up to a metre high can be produced by this method. This is also the method used for making polyisocyanurate blocks.

Phenolic foam

Closed cell phenolic foam is an exciting development which has come to the market over the past five years. Developed in the United Kingdom by BP Chemicals and Kooltherm, it has similar insulation properties to urethanes and isocyanurates, if not better, with the additional advantage of better flame spread characteristics and practically no smoke emission. This makes it ideal for the insulation of the inside of buildings where many people may be working. It is made by a continuous moulding process where each mould, approximately 2 m long × 1 m wide is passed through an oven, where the phenolformaldehyde resin is catalysed, the foam rises, and then cures. Again, panels of any length, width, and thickness can be produced using this method. The material is slightly more expensive than urethane foam and is meant for areas of high fire risk and/or areas where smoke generation must be kept to a minimum.

In deciding which rigid foam is most suitable for panel production, one must take into consideration the following characteristics:

- Tensile strength
- Thermal conductivity
- Moisture resistance
- Bond line between foam and facing materials
- Flame spread characteristics

Cork board (vegetable cork)

Cork is the outer bark of the cork oak tree, *Quercus suber*, grown principally in south-west European and Mediterranean areas, including Portugal, southern France, Morocco, Spain, Algeria, and Tunisia. These countries have between 4 and 5 million acres of cork forest, with an annual yield of 300 000 to 400 000 tonnes of bark. Portugal is the largest producer, with cork forests covering more than $1\frac{1}{2}$ million acres (equivalent to 10% of the total cultivated area of the country).

Cork bark and cork products consists of minute air cells having a a diameter of about 62 μm, with length just over twice the mean diameter. Each cell is a 14-sided polyhedron and is completely sealed from the next cell by a remarkably strong membrane consisting of five layers, with a total thickness of approximately 2 μm. In view of the very large number of cells in the cork bark, the material is highly resistant to the flow of heat because of the low thermal conductivity of the still air contained in the cells.

When the trees are some 20 years old the outer bark, which varies in thickness from 1 to 5 cm, is removed. The virgin cork so obtained is unsuitable for bottle stoppers, etc., but it can be broken into granules of various sizes to be used for cork composition products and insulation. Subsequent strippings occur at 8-year intervals until the trees have lived their life span, which may be anything up to 200 years. The third stripping, that is the layer nearest to the tree itself, is used for the manufacture of articles that are usually stamped out of the cork, such as bottle stoppers. It is the first stripping which is of interest for the manufacture of thermal insulation materials.

The granulated cork from the first stripping contains a natural resin which, under the action of heat and pressure, can be used to bind the granules firmly together. In the manufacture of thermal insulation, the granulated cork is packed into moulds and heated at a temperature of 300°C to release the resin which, on cooling, bonds the granules in a shape conforming to the moulds.

The resin released during the 'baking process' cannot be reused but the cork still retains its insulating properties and damaged or reject moulds can be broken up to yield 're-granulated' cork.

Granules of raw cork, exposed to high temperatures, expand slightly and at the same time lose their natural resin. With the addition of bitumen the granules can, under low heat, be moulded into pipe section and slabs for low-temperature insulation purposes.

Glass fibre

Glass fibres for thermal insulation purposes are produced from a molten mixture of sand, lime, and soda. The mixture is allowed to flow through tiny orifices in platinum bushes situated on the underside of the melting furnace, and then subjected to a blast of superheated steam. The mixture is at once broken into molten globules which in turn are drawn by air resistance into fibres of controlled length and diameter. Control is effected by the correct combination of melt temperature, steam temperature, and velocity. The fibres fall on to a conveyor belt, the speed of which governs the thickness of fibre blanket. The fibres can be used in this form for loose packing or in the production of mattress blankets, flexible pipe sections, and quilts. To produce a rigid slab, the fibres are bonded with thermosetting resins.

5.4   Insulation applications

The application of insulation to cold store structures depends primarily on customer preference. This normally depends on past experience, advice, or the customer's requirements in relation to cost, performance, and return on investment.

Structures may be internally or externally insulated. Internally insulated systems require wall and roof cladding.

Internal insulated systems with external wall and roof cladding

The steel structure (Figure 5.2) is designed to support not only the external cladding of the building, but also the insulated panels to the ceiling that may be suspended from the portals or clipped to the underside of either tied portals or lattice beams. The vertical steel work would normally incorporate the lightweight horizontal sheeting rails to give additional support to the wall panels (Figure 5.3a).

External insulation system

Insulated wall panels are fixed external to the steel structure, with the external face of the panels designed to be fully weatherproofed (Figure 5.3b).

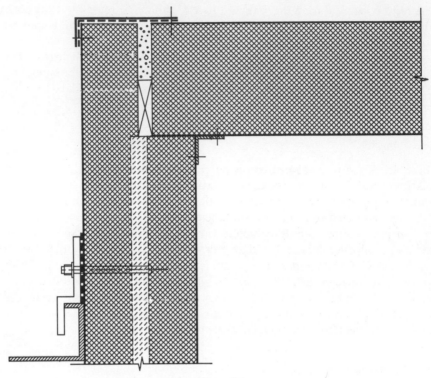

Figure 5.2

The insulation to the roof can be applied in two ways, either built-up from a profiled steel sheet secured to the steelwork with the insulation fitted in board form and a weatherproof cover over, or by laying insulated panels over the steel frame, sealing all joints, and then covering the whole area with a weatherproof membrane. In most cases the external membrane is then covered with white stone pebbles that reflect the heat from the sun and also act as a further support for the membrane itself.

The system of construction can also be used where static racking replaces the steel structure.

Internal insulation system with external roof and partial external wall cladding

This system, commonly known as a 'Dutch barn' construction, is designed to save costs. The insulated ceiling panels are secured to the steel portals but the wall panels are exposed to the elements, with the external cladding forming a valance around the periphery of the store protecting the external of the ceiling and the wall-to-ceiling joint from the weather (Figure 5.3c).

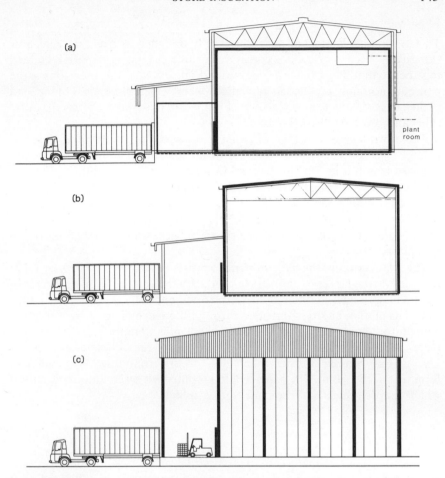

Figure 5.3 Types of insulation: (a) internal insulation systems with external wall and roof cladding; (b) external insulation system; and (c) internal insulation system with roof and partial external wall cladding.

The external joints of the wall panels are designed to give maximum weather protection and the external metal cladding will have a coating designed for protection against ultraviolet rays.

5.5 Types of cold store panel

The most common forms of cold store panels are:

- Continuous laminated panels
- Composite panels
- Foam injected panels
- Continuous foam panels

Continuous laminated panels

The continuous laminated panel, originally designed in Australia and now extensively used worldwide, consists of three layers of material bonded together so that they behave as one entity (Figure 5.4a). The outer skins of the panel are normally pre-coated galvanized metal which bears most of the load of the panel. the thicker central core of insulation material stabilizes the outer skins and prevents distortion under stress.

The advantage of this type of panel is the speed of production. Finished panel can be produced at a rate of 4 linear metres per minute, and to any required length using any of the rigid insulation materials mentioned above. Because of its superior bond line, it has stronger structural capabilities.

The continuous laminated panel, as with all composite panels, can be produced to any given thickness. This is a great advantage when it is considered that energy costs can be saved by increasing the thickness of the insulation.

The production of this type of panel consists of two coils of steel or GRP being fed through a series of rollers to produce a shallow profile to both faces of steel. This provides additional strength to the panel.

The insulation board is threaded through the machine, passing a series of temperature-controlled glue stations, where it is compressed between the metal skins which are stressed, and the curing of the adhesive takes place. During this process, further machining occurs, trimming the panel and forming the edge joint. The panel length is controlled by an automatic cut-off which shapes the ends of the panel into a rebate.

Composite panels

Composite panels normally consist of a board insulation, sandwiched between two steel facings (Figure 5.4b). Alternative facing materials may be used, such as GRP. The methods of manufacture can be either semi-automatic, where the panels are first laminated with the insulation being glued to the facing materials with a fast-drying adhesive, and then processed through a series of rollers ensuring that sufficient pressure is applied to both faces to form a positive bond, or by means of placing the panels after the laminating process into either a vacuum or hydraulic press to achieve maximum adhesion. This system is commonly used because of the flexibility in being able to use various forms of rigid insulation materials.

Foam-injected panels

These panels are formed by injecting polyurethane foam between the steel faces to produce a modular type of panel that can be used either with a dry joint or by incorporating a locking system (Figure 5.4c).

When utilizing a locking system for panels in industrial stores, one should consider the effects of the locking device in relation to the foam material taking into consideration the structural movement of the supporting steelwork.

Foam-injected panels (Figure 5.4d) have been used throughout the world for many years, mainly because of their superior thermal conductivity properties which allow the panel to be considerably thinner than those using alternative insulation materials.

Continuous foamed panels

These panels (Figure 5.4e) are produced on a continuous line. Two coils of steel are attached to the end of the machine and fed through a series of rollers. During the process, one side of the panel is constructed with a deep profile, with the other side left either flat or with a shallow profile.

As the two skins of steel are pulled through the machine, the two-part polyurethane foam is released onto the bottom coil of steel. With the injection of a blowing agent, it is agitated to allow the chemicals to mix to form a foam which rises and bonds to the upper metal skin. During the panel manufacturing process, the polyurethane continues to expand until it reaches its required density and the edges of the metal skin travel through a series of roll formers to produce the side joint of the panel. The panels are then cut to length by an automatic cut-off mechanism. This type of panel was produced primarily for the externally insulated panel system, where the profile steel forms the external finish of the store.

5.6 Present-day design criteria

- *Loads* The structure and the insulated panel system should be designed to resist the worst possible combination of forces considered below.
- *Manufacture* Stresses may be induced during manufacture due to the curing or ageing of the insulation and the handling of the panels. Procedures should be adopted to keep stresses within the design limits.
- *Erection* Care should be taken during erection, and procedures introduced to ensure that no abnormal stresses are created. Wind pressures can impose excessive stress on panels during erection, and considerable care should be taken to ensure that the incomplete structure is properly supported and that no undue pressure is placed on the supporting clips, etc.
- *Wind forces* The completed structure will be subjected to considerable wind forces. Consideration should be given at the design stage as to the best form of construction, taking into account the location of the store in relation to prevailing winds. Exposed panels on an exterior designed system, complete with the fixing system, should be designed to resist wind pressure in accordance with BS CP3 Chapter 5, part 2, 1972. The

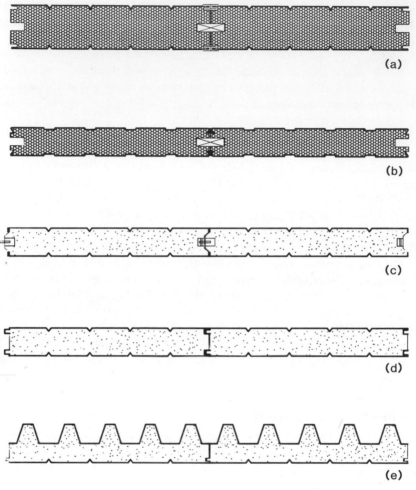

Figure 5.4 Types of panel: (a) continuous laminated panel; (b) composite panel; (c) foam injected panel with lock; (d) foam injected panel; and (e) continuous foam panel.

supporting structure should have no projections likely to damage the insulated structure when subjected to the effects of positive and negative wind pressures.

- *Dead and imposed loading* The structure, suspended insulated ceilings, and load-bearing walls should be capable of supporting the maximum combination of all loads. These should include

  –Dry weight of insulated panels
  –Weight of structure and fittings included to support insulated panels
  –Weight of coolers and fans under working conditions

−Weight of mechanical services
−External weatherproofing

On internally insulated structures, an allowance for miscellaneous items including loads due to occasional maintenance, should be the worst combination of:

−A uniformly distributed load of 0.23 kN/m^2 force applied to the complete area of the ceiling.
−A point load of 0.3 kN force applied to any isolated square metre of ceiling.

Excessive loading on the insulated panels in areas such as those occupied by control equipment should be avoided, and suitable walkways or catwalks, supported by the structural frame, should be provided above the panels. On external insulated structures, the imposed loading condition should be as described in BS 6399, Part 1, 1984.

5.7   Internal pressure relief valves

Modern cold stores utilizing the latest panel systems available are usually very effectively sealed. However, structural damage can be caused if consideration is not given to the changes in pressure within the store created by changes in internal temperatures.

Changes of internal pressure are generally dependent on the size of store or the use to which it is put, together with the type of defrost system which is introduced. These pressures can be safely relieved by the introduction of pressure relief valves strategically positioned in the wall panels at high level, opening and closing to ambient. Where stores are operating below 0°C, heater elements should be included within the valves.

The number and size of pressure relief valves for a particular store size is normally calculated by the refrigeration engineer, based on the size of the plant needed to maintain the operating temperature. To calculate the vent area required to keep the pressure difference within allowable limits, the following formula should be used:

$$A = 0.063Q/\sqrt{P(T+273)}$$

where A is the required venting area, in m^2, Q is the rate of heat production or extraction in the cold store, in kW, T is the cold store temperature, in °C, and P is the allowable pressure difference from store to outside, in N/m^2.

Although the required vent area does not depend directly on cold store volume, small stores are generally affected more than large stores for two reasons:

• Larger size of cooling or heating equipment in relation to store size
• Reduced air seepage into or out of the store, due to the level of sealing around the door gaskets

It is recommended that at least the following Q values should be considered when determining the maximum possible value of Q to be used in the formula:

- Maximum possible rate of heat extraction
- Maximum possible rate of heat input during defrost
- Maximum rate of heat input during any possible transient conditions, such as might arise after a defrost if cold air were blown through a hot-finned cooler

5.8   The importance of vapour seals

Walls and roofs

One of the most important aspects of cold store construction is ensuring that the insulated envelope is positively vapour sealed. Cold store users have over the years learned to their cost what happens to the structure if an effective vapour seal is not used on the warm side of the insulated panel. Water vapour will attempt to flow from a region of high vapour pressure area to a region of low vapour pressure, and if a barrier is not created to avoid the ingress of water vapour, the insulation will eventually absorb this moisture and, at low temperatures, turn to ice.

It therefore follows that the most important factors here are:

- Minimum permeability to water vapour
- Control of air vapour that may find its way through the joints of the panels

The panel itself may be impervious to water transmission, like an external steel-faced panel, but the joints between the panels must also be vapour sealed, as must all junctions and joints between wall and ceiling and between wall and floor.

The inside joint of the panel should have a higher permeability than the panel insulation so that in the event of intrusion of water vapour, it will pass through the joint and not permeate into the core material. It should also be noted that no material used in the panel joints on the store side should constitute a better barrier to vapour transmission than the vapour seal material itself. The water-vapour-tight joint should therefore have a permeability sufficient to prevent water or ice build-up in the insulation under the working conditions of the cold store.

The panel joint itself must be able to survive the differential movement between two panels and the sealant used must therefore have sufficient elasticity to tolerate this movement without breaking down. The main factors to consider are:

- The sealant must have a sufficiently low permeability to resist the vapour pressure differential that will exist

- The sealant must have good ageing characteristics
- The sealant, if a curing material, should not cure so quickly that the process has progressed too far before the whole joint has been sealed
- The adhesion qualities of the sealant are important. The materials should be capable of maintaining good adhesion to the panel surfaces in all working conditions and should also be compatible with any additional cappings that are applied
- In some cases it may be necessary for the sealant to be not only a vapour seal, but also water-tight. However, this should not be construed as being sufficient to protect an external cold store roof without additional weatherproofing
- The elasticity of the material should be such as to withstand the movements of panels, whether by live or dead loads; e.g. movements caused by thermal bowing and movements of the general steel structure to which the panels are fixed
- Wherever possible, stores should be designed to enable the maintenance engineer to inspect the vapour seal periodically. There should be sufficient room between the panels and the external cladding for this purpose. If this is not possible, the cladding should be easily removable to enable repairs to be made

It must be understood that in all applications of cold store construction, the effectiveness of the vapour seal depends on

- Use of the correct sealant
- Correct application of the sealant
- Protection of the sealant once the store is commissioned

Assuming that the sealant chosen meets all the physical and chemical requirements, we then come to the human element.

The application of the sealant must be properly supervised. It matters little if you are using the best material available on the market, if it is not applied correctly and in the right quantity.

The most vulnerable area of the cold store is its roof, due to pedestrian traffic. To protect the vapour seal at the joints on internal insulation systems, it is recommended that a metal cover strip is secured over the seal to prevent scuffing by pedestrians which might cause the sealant to be damaged.

It is unlikely, once the store is commissioned and the vapour seal tested, that the wall or floor areas will show any signs of vapour seal deterioration in the future. Any breakdown would normally be caused by mechanical damage. However, as mentioned earlier, the external surface of the insulated envelope should be accessible for periodic inspection.

The first signs of a vapour seal breakdown will be noticed on the internal joints of the store. For example, during warm and humid weather conditions with accompanying high dewpoint temperatures, the water vapour migrating through the insulation will be chilled to its dewpoint somewhere in the insulation and will condense to water. In the case of store areas above freezing the insulation will remain wet; in a freezer ice will form, and build up until it is visible on the internal joint.

At the first signs of ice or snow on the joint, the external vapour seal should be checked and repaired and the ice or snow removed. If these conditions are allowed to continue without repair the value of the insulation and the structural stability of the panel will gradually decrease until it is finally destroyed. Whilst some insulation materials are more impervious than others, it should be recognized that with an external steel face to the panel, the panel itself is impervious to water vapour transmission. It follows that the only areas susceptible to the transmission of vapour are the joints. It is generally acknowledged there is no known way to guarantee the long-term performance of the jointing seals completely, so consideration should be given both to the type of insulation used and to the design of the joint.

The joint must be thermally continuous, and should have a higher permeability than the insulation core. This is because the water vapour will take the line of least resistance and if the joint material is as good or better than the core insulation from the point of view of water vapour permeability, then the entering vapour will permeate into the core instead of through the joint.

The ideal joint is therefore one with a tight but dry straight-through joint, which will give adequate thermal integrity and the passage of least resistance to any intruding water vapour. That is not to say that stepped joints or insulated tongues cannot be used, as long as they are applied dry. The structural stability of the panel must be considered at all times.

Any water vapour that intrudes through this joint will continue to migrate until it reaches the refrigeration coils, where it is deposited in the form of ice provided that its path is not blocked by any seal having lesser permeability than the panel insulation.

Floors

We have now dealt with the sealants used for vapour sealing the joints to the walls and ceiling of the cold store, but we must appreciate that to continue the vapour seal under the floor insulation is of equal importance. The complete thermal envelope of the cold store includes the floor insulation and the insulation must be continuous on all sides, so it follows that the vapour barrier must also have all-round continuity.

We will now examine the continuity of this vapour barrier under the floor insulation. To achieve an effective seal to the floor, 1200 gauge black polythene laid over a smooth prepared cement floor is the best proven material to replace the liquid vapour seals that were used in conventional-style buildings.

The polythene is laid continuously over the full floor area before applying the floor insulation. All joints should be lapped and bonded together. To

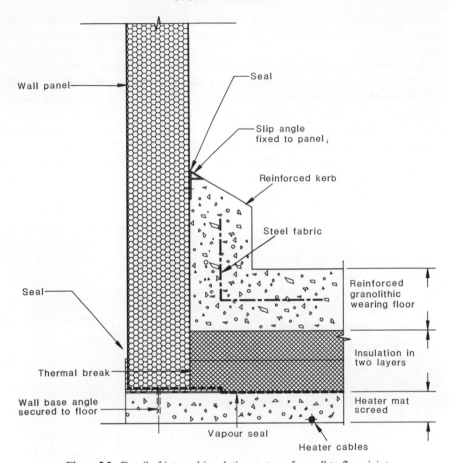

Figure 5.5 Detail of internal insulation systems for wall-to-floor joint.

complete the envelope, the polythene must continue under the panels to be bonded to the metal wall support plates (Figure 5.5).

Care should be taken when laying the floor vapour seal that there are no projections or loose stones in the sub-floor that could penetrate the vapour barrier. However, if a recognized extruded polystyrene insulation is laid over the vapour seal, the risk of any moisture from the sub-floor permeating through the floor structure is minimized.

5.9 Thermal bowing

Thermal bowing can be defined as the stress caused by temperature variation between the internal metal face of the insulated panel and the external face. The degree of thermal bowing will depend primarily on the location of the insulated enclosure. The maximum store temperature depends on whether the panel is exposed directly to the sun or is protected by an uninsulated external cladding.

It should be noted that when panels are installed on the external surface of the steelwork or exposed to the elements, consideration must be given to the fabric and the colour of the exterior coating. The skin temperature of a panel subjected to ultraviolet radiation can be greatly reduced by utilizing a particular colour. At the design stage the correct colour and material most suitable for the installation should be established from the steel producers.

Ceiling panels, although subjected to thermal bowing, are generally not affected to the same degree as the wall panels. Modern methods of

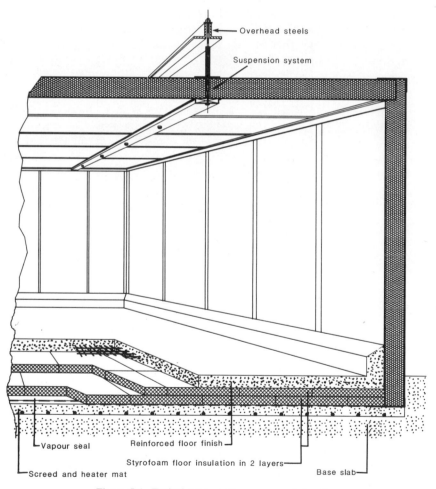

Figure 5.6 Typical store section — suspended ceiling.

construction usually allow the ceiling panel to be supported between the steel structures of the building (Figure 5.6). The unsupported span of the panel will depend on the thickness of the insulation, the gauge of the steel fabric, and the bond line. The temperature difference between the internal and external steel faces will cause the panel to bow upwards, but this is normally counteracted by the weight of the panel. In some cases where panels are designed to span over support steels, secondary stresses may be induced and precautions should be taken in the method of fixing and the size of the panel used.

It should be recognised that the effect of thermal bowing on wall panels can be serious, and every effort should be made to minimize the risk of structural damage. It is therefore important to ensure that the method of fixing is adequate to accommodate the forces created by the differing movements of the facing materials.

Panels are generally produced to extend the full height of the store, and are normally supported at intervals by fixing to horizontal steel rails. However, if the conditions are such that by fixing through the panels the thermal bowing affect cannot be properly restrained, then consideration should be given to the panel design and the need to relieve the steel facing midway in the height of the panel.

To minimize the effect of thermal bowing on cold stores with internal insulated systems (Figure 5.3) a means of ventilation should be introduced. Where there is a void between the top of the insulated panel and the external roof, the area should be ventilated to reduce solar heat gain. The ventilation system can be installed either by vents or louvres in the gable walls, coupled with the ridge vents, or by introducing a fan to create forced ventilation over the ceiling area. This also applies to the void between the wall panels and external cladding.

5.10 Construction methods

It is important before commencing construction of the insulated enclosure to establish the site conditions. As with all methods of construction the following factors should be observed:

- *Accessibility to site* Access to site by established roadways or temporary roads should be observed, and decisions made as to the type of vehicles and plant most suitable for the transportation of panels and ancillaries to the designated areas.
- *Ground conditions* Are the areas adjacent to the steel structure properly prepared to support scaffolding, mobile towers, scissor lifts, etc. that are needed for the installation of the panels? Checks should be made that the designated areas for the storing of panels and materials are prepared and level to avoid any on-site damage.
- *Availability of services* Before commencing work it must be established

when main services, such as electrics and water, may be installed, or whether a portable generator will be available. It is also necessary to establish a location for the site hut and canteen facilities for workers on site. Arrangements should be made for a telephone to be available at all times. It is important to maintain a high level of communication between the site office and headquarters, and with suppliers.

- *Safe working conditions* Safe working conditions must be recognized and understood by both management and workers on site. Safety equipment should be available, and conditions on-site should comply with the Health and Safety at Work Act 1984.

Internal insulated systems with external cladding to walls and roof

This sort of insulation system has been described in section 5.4. Before planning the erection sequence of the panels, it is important to establish that materials will be available from the supply factory at all times, to ensure continuity of work, and to support the site programme.

In this method of construction, it is normal to commence construction once the heater mat and sand and cement screed have been installed, and the external roof cladding complete. As the panels are installed on the inside face of the steel structure, the external cladding can be finished at the same time.

Whenever possible, wall and ceiling panels should be installed in such a way as to allow the refrigeration engineer to install the coolers. For example, working away from the plant room, assuming that this is located adjacent to the coolers, wall panels will be erected in a U-shape: that is, back wall first, then side walls and ceiling panels installed together. This reduces the site time and allows the floor insulation to be laid also, making it possible for all allied trades to follow on behind.

To support the wall panels, a heavy-duty galvanized steel angle is first fixed to the concrete floor, level with the internal face of the main steel structure. This angle is bedded in a mastic solution and extended approximately 150 mm beyond the internal face line of the panel (see Figure 5.4). The angle is secured well within the area designated for the periphery of the heater mat. It is of paramount importance that this area should be checked thoroughly before any penetration into the concrete is made that could possibly cause damage to the heater elements or glycol system.

Before installing the wall panels, a strip of black polythene vapour seal is laid to cover the base of the angle and to extend into the room. This will form the continuity of the vapour barrier at a later stage. The wall panels can now be installed making sure that the panels sit firmly on the black polythene vapour seal and are sealed to the vertical upstand of the angle.

The first panels to be installed will be a wall corner. The joints will be either rebated together, or shaped to form a mitre. It is important at all times to ensure that these starting panels are perpendicular and that the insulated joint is well fitted. Each corner panel is secured to the horizontal steel supports and also fastened to the base section and sealed with mastic.

Once the first corner section has been positioned, the back wall adjacent to the plant room can be installed. Whatever type of panel has been selected, it is important that the joints between the panels fit tightly and the vapour seal properly applied.

One method of ensuring a vertical tight joint is to remove the bottom 150 mm of internal steel facing once the panel is secure. This section should, in any event, be pre-cut before leaving the factory (Figure 5.5). With the top section of the wall panel cut into a rebate (Figure 5.7) exposing the insulation and with the bottom 150 mm exposed, it is then possible to see at all times that, if the joint of the panel is tight at the base and at the top, the complete vertical joint, whatever height, will always be a good fit. As each wall panel is erected, the vapour seal to the joints is also applied.

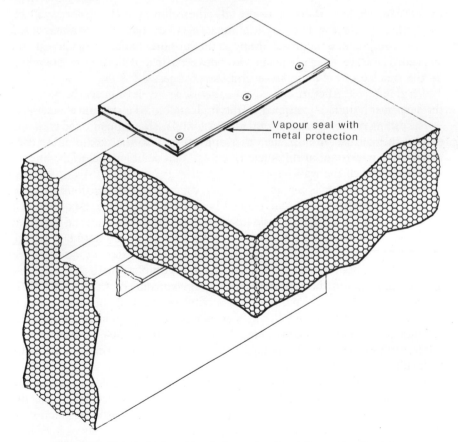

Vapour seal with metal protection

Figure 5.7 Detail of internal insulation system for wall-to-wall ceiling joint.

One important factor that should never be ignored is the fixing of the panel to the steel rails immediately the panel is erected. This ensures not only the safety of those working in the area, but, in the event that the external cladding is incomplete, it also secures the wall panels against high winds. Once the back wall has been installed and the next corner section complete, it is then necessary to prepare the ceiling suspension units. These can be installed by one gang while the other erectors are installing both the side walls. As each suspension unit is fitted, the ceiling panels can be installed. This allows for the wall and ceiling of the store to be fitted simultaneously.

There are various methods of ceiling construction, depending on the type of steel structure that is used. The main structural roof members are normally either:

- Steel or concrete portals
- Tied portals
- Lattice beam construction

In the case of steel or concrete portals, the ceiling panel supports will be suspended by hangers. The length of the hangers depends on the angle of the roof and will be cut or adjusted on site as the support channels are aligned and levelled. These are normally positioned between 6 m and 7 m apart, depending on the spacing of the main structural steel frames.

When the steel structure consists of either tied portals or lattice beams, then the suspension units are clipped to the underside of the structural steel ties.

The ceiling panels are next installed. The first panels are positioned with one end on the suspension unit, the other end fitting into the rebate of the wall panels. It is important at this stage to ensure that ceiling panel joints line up with the joints of the wall panels.

As the ceiling panels are installed, the vapour seal to the joints is also applied, as with the walls. At this stage, where the ceiling panels rest on the wall panels, these joints are left dry and filled and vapour sealed once the insulated envelope is complete. It is important that these joints are left until last as they require special attention. This detail will be described later.

The first part of the erection sequence is now complete, with the back wall and the first bay of the wall and ceiling panels installed. At this stage it is important that the internal corner angle between the wall and ceiling panels are completely secured, together with the fixing of the suspension units to the ceiling panels. In both cases the angles and suspension units are fitted dry. In the case of the suspension units, a gap will be left between the ends of the panel to be filled and sealed later.

Once the first bay has been completed, it is then possible for the ceiling evaporators to be suspended and the necessary pipework completed in the area. The erection towers and lifts can then be moved to enable the erection of the next two bays of panels to be completed.

Dependent on the site programme, it is now possible at this stage to commence the installation of the door frames and floor insulation. It is important to remember that the floor insulation and floor vapour seal can only be laid once all services are completed to the underside of the ceiling panels, i.e. no scaffold towers, etc. should be allowed on the floor insulation until the internal floor finish has been laid in the area and cured sufficiently to accept either mechanical or pedestrian traffic. The remaining panels can now be completed to form the insulated enclosure.

Following the installation of the wall and ceiling panels, the next stage of erection will include the completion of the floor insulation and vapour seal. As the vapour seal is laid over the floor, ensuring that all joints are lapped and bonded together and also bonded to the vapour seal projecting from beneath the wall panels, the insulation can also be laid.

Internal insulation with external roof and partial external cladding

This system of construction is designed generally as for the internal insulated system, with the exterior walls forming a weatherproof surface. The wall panels are therefore designed to overlap the base concrete to form a watershed, as for the external system of construction.

External insulation systems

The principle of the insulated enclosure remains the same as for the internal system. Panel joints have to be properly sealed with the insulation and vapour seal forming a complete envelope. With this system, the wall panels are erected on the external of the steel frame and are constructed to form a weatherproof unit. The roof insulation can either be a built-up system of insulated boards, or can be constructed using prefabricated panels with a weatherproof cover applied.

The external insulation system can also be applied to pre-erected racking that would be constructed in such a way as not only to replace the main structural steel, but also to reduce the overall cost of the cold store considerably.

As with the internal system, the wall panels are erected first ensuring that all the panel-to-panel joints, together with corner joints, are not only vapour sealed but also weatherproofed.

The roof structure, if constructed on a built-up system, normally consists of sheets of profile steel fixed to the top of the structural steel with the insulated boards laid above this. This is followed by sheets of compressed board that will form the base for the vapour seal and weatherproof roof. Where insulated panels are fitted, these replace the material used for a built-up structure.

In this form of construction, the structural steel is designed to allow for the roof insulation to be laid to falls. The wall panels forming the gable ends extend above the roof insulation creating a parapet to which the vapour seal and weatherproofing can be dressed to give continuity of the seal and insulated envelope. Many types of material are used for the weatherproofing of the external roof system. The most common material is butyl rubber stretched in rolls over the complete roof area with the joints lapped and fused together. The material is double lapped over the parapet and secured with a weatherproof capping to the wall panel, completing the vapour seal. Layers of mineral felt can also be used in conjunction with the butyl rubber to give additional weather protection. The exterior of the roof is then covered with a ballast of stone pebbles.

Floor insulation

The most acceptable form of insulation used in floor construction in the UK is extruded polystyrene foam slabs. Because of its superior compressive strength and high level of moisture resistance, it makes an ideal material for this application.

The extruded polystyrene is produced in approximately 2400 × 600 mm slabs and in thicknesses of 50 mm, 75 mm and 100 mm. Unlike the extruded polystyrene used in the construction of wall panels, this material is left with a skin on both surfaces, giving additional moisture resistance and also a more durable surface for floor applications.

Once the floor vapour seal has been applied, the boards of extruded polystyrene are then laid in two layers with all joints left dry and staggered to ensure no straight-through joints.

The first layer is laid in brick fashion, with the peripheral joint cut and fitted tight to the insulation that is exposed at the base of the wall panel. Again, working away from the back plant room wall once several bays of the first layer of insulation have been applied, it is then possible to continue with the second layer. The second layer should be applied like the first, with the exception that the boards should be applied at right angles to the first layer, i.e. first layer working away from the back wall and the next layer across the room.

Normally, because of the consistent flatness and rigidity of this type of insulation, it is not necessary to fix the layers together. However, should there be any slight undulation in the concrete base slab causing the insulation to stand proud of the adjoining slab, the slabs can be secured by fixing the top layer to the bottom with beechwood skewers. In any event, it is good practice to examine the completed floor insulation and apply skewers to any area of insulation that may not be fitted tightly.

During the application of the floor insulation, it is possible to work simultaneously on the exterior of the cold store and to fit the door frames.

At this stage, all panel-to-panel joints will have been completed with the vapour seal applied during the installation. Work remaining to be completed includes the finishing of the junction between wall and ceiling panels, and the joint left between ends of the panels where they rest on the suspension units. The easiest and most efficient way of filling these joints is to apply polyurethane foam. Slab insulation can be fitted, but this is time-consuming and does not necessarily guarantee the integrity of the joint.

It will be recalled that the internal wall-to-ceiling corner angles were fitted dry. It is therefore necessary when applying the foam to ensure that it does not run out at this joint at the pre-expansion stage. The best method of avoiding this is to cut strips of insulation approximately 50 mm to fit tightly into the bottom of this joint. The polyurethane foam can then either be mixed and poured into the joint or applied direct from an aerosol canister. The foam should be allowed to rise 50 mm above the joint to allow for cutting back to the exterior level. This also creates a good surface for applying the vapour seal. A self-adhesive bituthene strip or mastic can then be applied over the joint before fitting the metal protective angle. Considerable care should be taken to ensure that this joint is properly fitted and that the vapour seal is applied in the correct manner, with the vapour seal and metal strip extending at least 50 mm onto the roof panel and secured with retaining rivets. The joint above the suspension unit should be filled and sealed in a similar manner, and the vapour seal should be protected with a metal strip.

Before leaving the exterior of the roof, all joints should be checked and additionally sealed if required. Special attention should be directed to the support detail of the suspension units. If steel hangers are used, then these should be insulated to a height of approximately 300 mm above the roof panel.

If the external wall-to-wall corners have not been completed at this stage, they should now be fitted with the appropriate vapour seal and protective corner angle.

During the installation of the wall panels, apertures would have been left to accommodate the doors. It is now possible to fit the doorframes into these apertures, leaving the hanging of the doors until the internal floor finish has been completed. It is important to ensure that the integrity of the insulated envelope is maintained during the installation of the frames, and that the reveals are fitted in such a way as to avoid cold tracking and the consequent formation of moisture or ice.

The internal floor surface can now be laid, and then all doors can be installed.

5.11 Underfloor heating

Any building designed to operate at temperatures which can cause the formation of ice under the floor must be protected. It can be safely assumed that any room operating at below 0°C will require some degree of protection. This is dependent on a number of factors such as location (whether internal to the existing building complex or external); geographical location; actual room temperatures; type of operation, etc.

Four basic types of system are commonly used to protect the freezer floors from frost heave:

- Circulated glycol system
- Low-voltage electric heater mat
- Structural raised base slab
- Vent systems

Whichever system is preferred, it is important to provide some means of monitoring the underfloor temperature so that a potentially damaging situation can be averted.

Circulated glycol system

The circulated glycol system is the one most commonly used where the floor area exceeds $1500 \, m^2$. The system comprises of a grid of plastic tubing laid in loops and connected to a header pipe through which a warmed solution of glycol is circulated (Figure 5.8). The circuits should preferably be in continuous runs with no joints under the floor. The tubing normally consists of 25 mm internal diameter polyethylene, which is supplied in approximately 300 m rolls. The tubing is held in position with wire and fixed to the sub-floor. It is then covered with a 75 mm sand and cement screed.

It is important that the tubing is tested prior to the pouring of the screed and that the test pressure be maintained and observed during the entire construction process.

The heat to warm the glycol is normally obtained from the central refrigeration system where a discharge gas heat exchanger can be utilized to reclaim heat. A pump is then used to circulate the glycol solution.

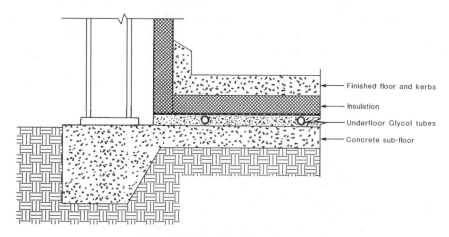

Figure 5.8 Circulated glycol system.

As with all mechanical underfloor heating systems, a monitoring system should be introduced to ensure that the temperature of the floor below the insulation stays above freezing (approximately $+5°C$).

The glycol system is an economical form of underfloor heating once the initial cost of installation has been absorbed. While it requires some space for the heat exchanger, pump, monitoring equipment, etc., the header and tube connections can be placed against the wall and easily protected.

Low-voltage electric heater mat

This system consists of a grid of electrical heater elements (Figure 5.9). The cable consists of stainless steel wires protected by a heavy-duty PVC covering. The grid comprising of individual circuits extending into a busbar chamber, which is then controlled by a tapped transformer from the main electrical supply.

Once the grid of wires has been secured to the sub-floor, all circuits should then be tested. The wires are then further protected by a 50 mm sand and cement screed. It is of great importance that once the screed has been laid, all circuits are again tested to ensure that no damage has occurred during the laying of the protective screed.

When setting out the grid for the heater cables, it is important to ensure that the peripheral wires are positioned well inside the line of the wall panels, to prevent any possible damage caused by the fitting of the wall base channel.

The low-voltage heater mat system is a good choice for stores with a floor area of less than $1500\,\text{m}^2$.

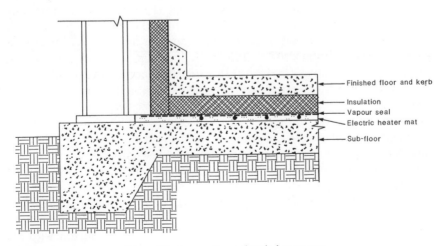

Figure 5.9 Low-voltage electric heater mat.

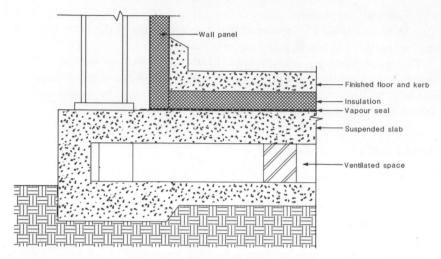

Figure 5.10   Structural raised base slab.

Structural raised base slab

Mechanical underfloor heating can be ignored if the cold store base is constructed on concrete or dwarf walls, allowing for a ventilated space between the slab and ground level through which ambient air can pass (Figure 5.10). This system is normally applicable when the site needs to be piled to support the cold store construction.

Vent systems

The gravity vent system, which is more common in the USA than in the UK, is a series of tubes placed under the sub-floor slab (Figure 5.11). The tubes are placed on a slope of approximately 1 in 50 to induce drafts; that is, as the air cools it falls to the lower end of the tube and warm air replaces it. The tubing is generally 100 to 150 mm PVC positioned at approximately 1200 mm centres.

With this system, however, there are a variety of theories as to the best depth for the tubes below the internal wearing surface, and also the slope of the tubes.

The advantage of this system is that it is simple and can be normally constructed by any contractor. The disadvantage is that it requires consistent maintenance to ensure that the vents remain open, as they are subject to blockage by vermin, dirt, and ice. The gravity vent system is certainly not recommended for larger cold stores but can work well in very small stores.

The forced draft vent system is supplementary to the gravity vent system, and should be used only in very small stores. The problem of not having sufficient air and heat flow in the gravity system can be solved with the addition of a *properly sized blower* located in a warm area. The tubes should be

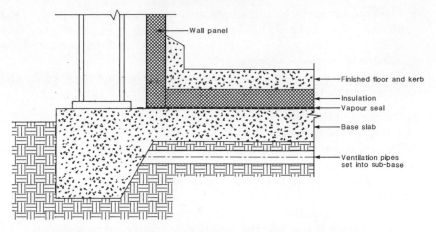

Figure 5.11 Gravity vent systems or forced draft vent system.

sloped in the direction of the air flow to allow condensate to drain out. This system, as with the gravity system, can only be effective with constant maintenance ensuring that the tubes are always kept clear.

5.12 Specification for internal floor finishes

Generally the most commonly accepted internal floor finish to cold stores is reinforced (granolithic) concrete. However, an alternative to this is a reinforced (monolithic) concrete; the only advantage is a marginal saving on the overall cost of the store.

Whichever floor wearing surface is chosen, it must:

- Possess a hardwearing and low-dusting surface
- Be capable of withstanding intermittent and rapid changes in temperature without cracking
- Be level within the tolerances necessitated by a racking installation and mechanical truck operations

The wearing floor shall be capable of transmitting onto the insulation and the base slab without undue stresses, the maximum load and point loads from racking installation, mezzanine floors, FLTs, etc.

In the case of mobile racking, heavy-duty EPs should be fitted below the concrete directly under the tracks that support the wheels.

The specification for the internal wearing floor shall comply with the following requirements:

- The general requirements of CP110, Part 1: 1972.
- The general requirements of the recommendations for concrete floors published by the Cement & Concrete Association.

- The aggregates shall be granite or satisfy the requirements of BS 882 1983 for heavy-duty concrete finishes. Concrete shall be made with specially selected aggregates of a hardness, surface texture, and particle shape suitable for use as a wearing surface.
- The concrete mix shall have a minimum cement content of 330 kg/m^3 and a minimum crushing strength of 35 N/mm^2 after 28 days.

Construction joints should be arranged beneath racks where fitted and, as far as possible, out of the way of gangways used by FLTs. Where this is not possible, precautions should be taken to provide a joint with similar resistance to wear as the remaining area of the floor.

A kerb or similar protection should be constructed around the perimeter of the wearing floor. The kerb should be structurally connected to the floor by allowing the metal reinforcement to be returned at right angles into the kerb. The gap between the kerb and the insulated panels should be sealed after the store operating temperature has been reached, or be covered with a cover strip that allows for the movement of the kerb during the temperature pull-down.

The wearing floor shall be allowed sufficient time to cure prior to temperature reduction or loading. Care should be taken to prevent expansion of the wearing surface prior to cooling, as expansion joints are generally not provided.

Sealant finishes are sometimes applied to the wearing floor to reduce dusting. The sealant should be applied in accordance with the maker's instruction and should be of a type which does not cause lingering taint.

Cooling the store down to temperature

Once the internal wearing floor is cured, a cooling-down period should be observed. Cooling should be slow to prevent the floor cracking, and carefully controlled as the temperature approaches 0°C, to promote drying prior to cooling below 0°C. A typical cooling programme is shown in Table 5.1

5.13   Insulated doors

All insulated doors to the cold store should be designed with insulation equivalent in value to that of the wall insulation and, in the case of large doors,

Table 5.1 Cooling the store to temperature: a typical schedule.

Day 1	16°C	Day 7	−4°C
Day 2	5°C	Day 8	−8°C
Day 3	1°C	Day 9	−12°C
Day 4	1°C	Day 10	−18°C
Day 5	1°C	Day 11	−20°C
Day 6	−2°C	Day 12	−29°C

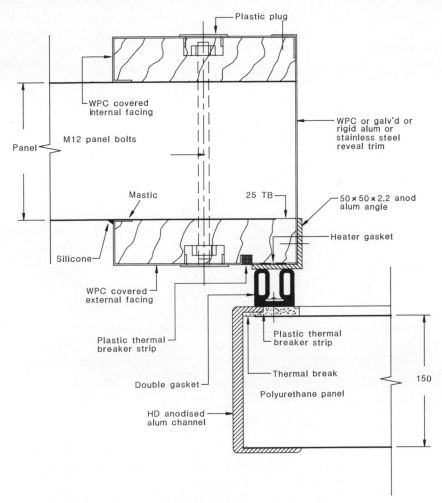

Figure 5.12

they should be constructed in such a way as to minimize the effect of bowing caused by the temperature difference between the internal and external surface of the door.

Doors should be constructed using a rigid frame or a heavy-duty metal channel and should incorporate a thermal break between the internal and external temperatures (Figures 5.12 and 5.13).

Doorframes should be made of material not only capable of supporting the door but also constructed to ensure the integrity of the walls to which the frames are fitted.

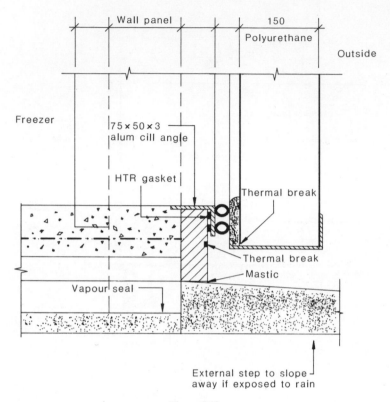

Figure 5.13

The most common form of door insulation is polyurethane. As this insulant has a superior thermal conductivity value, it allows the doors to be of minimum thickness, thus reducing their overall weight.

The doors should incorporate a heavy-duty gasket to form an effective seal. In the case of sliding doors, the gaskets should be resilient enough to absorb the pressure of closing and also any drag caused by the sliding operation.

There are various types of cold store doors:

- Main sliding doors
- Main automatic sliding doors
- Horizontal automatic bi-parting doors
- Vertical slide doors (manual or automatic)
- Hinged personnel and fire escape doors

Wherever possible all doors should be fitted on the exterior warm side of the

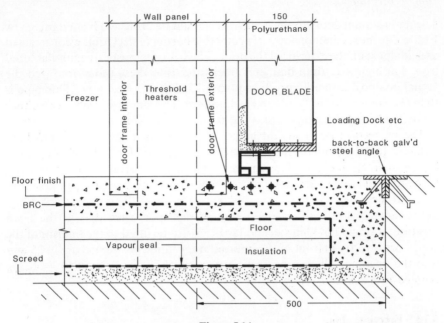

Figure 5.14

cold store. Where stores are operating below 0°C all doors or doorframes should be provided with heater elements to prevent sticking due to ice build-up (Figure 5.12).

Heater elements should be transformed down from the main supply, be free of fixings, and be insulated in accordance with the manufacturer's recommendations.

Threshold heaters should also be installed in the wearing surface of the floor and have separate connection to that of the doorframe heaters (Figure 5.14).

Doors fitted on the outside of a weatherproof wall should have a metal canopy fitted over the door head to protect the gasket and heaters from weather conditions.

All door furniture should be corrosion-resistant, and large enough to ensure the easy operation of the doors.

Automatic sliding doors are not acceptable as a means of escape unless they have a manual override and can be opened manually in the event of power failure. The locks to these doors should isolate the drive mechanism when the door is locked. A safety mechanism should also be fitted to avoid injury or damage to produce should the door accidentally close. All doors required for means of escape should be easily and immediately operable from the inside at all times, and should be identifiable to all operatives.

Door protection

The main sliding doors to the cold store should be protected from damage by FLTs. The most common form of protective barrier is steel tubing constructed as a goalpost on the exterior of the door. The steel goalposts are manufactured from 1 m high × 150 mm diameter steel tube. Into these tubes are fitted the actual goalpost assembly produced from 75–100 mm steel pipe. This pipe is then cemented into the 1 m high steel bollard. The whole assembly can either be bolted to the floor or set into the main concrete sub-base.

The protection posts should be positioned so that the bottom 1 m high bollards, together with the head of the goalposts, extend into the clear opening by approximately 25 mm. This ensures that the FLT driver can use these as a guide to avoid hitting the door reveals when the door is in an open position.

A further protective barrier can be extended from the goalpost assembly, approximately 1 m high, to the extent of the door when it is in the open position. A single 1 m high steel bollard can also be fitted to the interior of the cold store directly in front of the internal door frame positioned again 25 mm into the opening. All steel protection barriers should be identified by being painted in black and yellow stripes.

5.14 Erection plant

It is important to ensure that during the erection of the cold store, the correct equipment is used and that it is designed and erected to conform to the appropriate safety standards. The general erection equipment should be:

- Rough terrain FLT
- Smooth-wheeled FLT
- Mobile scaffold towers
- Heavy-duty mechanical scissor lifts

The rough terrain FLT should be capable of off-loading and transporting panels and ancillaries over rough ground to the designated areas.

The smooth-wheeled FLT is for use inside the building and for travelling over a smooth floor, with maximum load, without damaging the heater mat protective screed.

Dependent on the size of the store to be built, mobile towers should be installed to enable speedy erection of the wall panels. These should be erected by the supplier and should be designed, when fully erected, to comply with the appropriate safety standards. Outriggers should be fitted at all times to ensure maximum stability of the tower when erected to full height. The working platform of the mobile towers should be fitted on all sides with handrails positioned above the waist height of the tallest man working on the platform.

The use of mechanical scissor lifts, with controls from the working platform, for fitting ceiling panels will reduce the erection time considerably. These lifts are fitted with all the appropriate safety features. However, operatives should be properly supervised in the handling of this equipment and should keep well clear of all moving parts during the ascending and descending of the lift.

On external insulated stores it may be necessary to erect fixed scaffolding. Men working near the edge of the roof should at all times be fitted with safety harnesses.

All plant and erection equipment used on the installation of cold stores should comply with the Safety at Work Act 1984, and the erection teams should familiarize themselves with all aspects of safety on site.

5.15 Technical data

The insulation thicknesses shown in Figure 5.15 are based on ambient temperatures of + 20°C. The graph does not take into consideration structural strength requirements of insulated panels, which require additional thickness

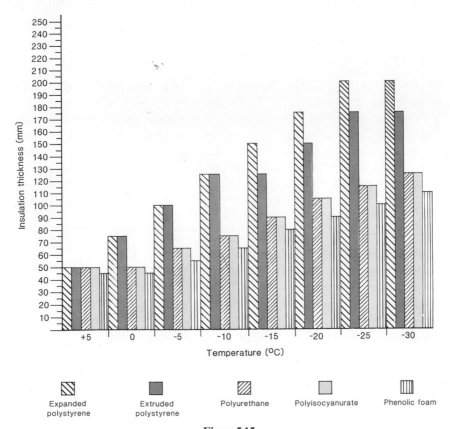

Figure 5.15

Table 5.2   Physical properties of insulation materials.

Physical property	Expanded Poly-styrene	Extruded Poly-styrene	Poly-urethane	Polyiso-cyanurate	Phenolic foam
Maximum thermal conductivity at 10°C, W/(mK) Aged value (30 days)	0.035	0.029	0.021	0.021	0.020
Maximum compressive strength or compressive stress at 10% strain, KPa	110	300	100	100	100
Maximum water vapour permeability, ng/(Pasm)	5.0	1.7	5.5	8.5	5.5
Apparent water absorption maximum, % (vol)	4	0.2	4	4	7.5
Fire characteristic (minimum acceptable performance level when incorporated in a panel under BS 476 PART 7 & 8)	CLASS 1	CLASS 1	CLASS 1	CLASS 1	CLASS 1

to provide a span performance that can cope with the worst possible combination of forces.

British and European refrigeration experience shows that the most economic thickness of insulation permits a mean maximum heat flow through the total surface area of walls, floor and roof of a cold store, approximating $8 \, \text{kcal/m}^2/\text{h}$. American practice over the last ten years has demonstrated that considerable cost savings can be made both in capital cost of the refrigeration plant and energy saving by increasing the thickness of insulation.

Some physical properties of insulation materials are given in Table 5.2.

5.16   Regulations and standards

UK requirements

Lloyds Register of Shipping   Within the UK, Lloyds Register of Shipping Type Approval has been accepted as the recognized standard for the manufacture and construction of cold store insulated panels. Cold stores required to be built to Lloyds Standards must conform to *Rules and Regulations for the Classification of Refrigerated Stores, Container Terminals and Process Plants*, January 1988 edition.

The development of Lloyds Type Approval has been considerable, especially during the last ten years, during which improved standards have been bought about by dramatically increased costs involved in the insurance of cold store buildings and by adaptations to modern distribution and storage needs. In addition, the value of goods stored in cold stores has increased considerably and any failure on the part of the insulated envelope greatly increases the insurers' liability. The major areas of interest to the assessors are as follows:

- *Panel specification* This area includes a physical assessment of the panel components, including assessing the value of the properties of components after manufacture. Within this category are the mid-span loading and deflection details, as well as the long-term structural effect of load on both horizontal and vertically installed panels.
- *Construction details* Included within the area of construction details are: jointing systems; door leaf and frame construction; internal and external vertical corner detail; ceiling supports; external wall detail; wall-to-ceiling and wall-to-floor jointing arrangements.
- *Ancillary construction details* The interest of the surveyor has been expanded to include the assessment of ancillary components related to the construction of the insulated envelope. The type and specification of mastics, as well as edge jointing details, are now part of the investigative process. The expansion of the surveyor's investigation to such ancillary items is a departure from the traditional concentration on manufacturing technique, which was a standard initially introduced by Lloyds Register of Shipping when Type Approval was first developed. Lloyds Refrigerated Store Certificate, known as the *Maltese Cross*, will be assigned when the relevant parts of the cold store have been constructed, installed and tested under special survey, and found to be in accordance with the rules.

British Standards Institution The BSI is the independent national body for the preparation of British Standards. It is the UK member of the International Standards Organization and UK sponsor for the British National Committee of the International Electrotechnical Commission.

In preparing a British Standard Specification, committees are formed from institutions, advisory boards, professional associations and government bodies, which, because of their technical expertise, are responsible for the preparation of standards.

Consultants and architects familiarize themselves with all appropriate standards when designing a cold store complex, and whenever possible manufacturers and contractors will conform to whatever standard is applicable.

The Institute of Refrigeration The Institute of Refrigeration plays a major part in both refrigeration and insulation services, and through its membership of engineers and allied trades has produced that *Code of Practice for the Design and Construction of Cold Store Envelopes, Incorporating Pre-Fabricated Insulated Panels,* 1986 edition. This incorporates the appropriate legislation and also the British Standards Codes of Practice for the construction of cold stores.

European requirements

The development of building codes for the construction of cold store buildings in Europe was initially absorbed in the general building industry construction requirements. In many countries building material tests have been utilized to assess insulated panels. The German 'funnel-fire test', for instance, relates to a variety of insulated materials used in general construction and not necessarily specifically to the insulated panel technology available for cold store construction in Europe today. For this reason, the Lloyds Register of Shipping Type Approval is rapidly becoming a standard throughout Europe, because it assesses panel strength, vapour seal, and insulation values under low-temperature conditions.

In the continual development of building codes for the cold storage industry in Europe, it is likely that Lloyds Register of Shipping Type Approval will become the norm, and will be accepted and recognized on a broad basis, especially in 1992 when Europe will move rapidly into one market.

US requirements

There are no standards of construction in the USA equivalent to those published in the United Kingdom. System design and selection is left entirely to the architect, in contrast with the UK where standards are fairly well set in terms of design criteria, and do not allow much in the way of variation.

The National Association of Cold Storage Contractors is preparing a Standards and Practices Manual related to cold store construction. These standards will be for informational purposes only and will not form the basis of a government regulatory document. The standards make the assumption that a competent designer has been engaged, who will select the system style and components best suited to a particular project. On that assumption the standards will then provide the following general requirement:

- The ability of the insulation contractor to provide the appropriate materials and system to satisfy the architect.
- Requirements by the architect that the construction will conform to design and that the insulation and components will form an envelope relative to the trueness of planes, and will withstand the wind forces imposed on the completed installation, together with dead and imposed loads created by mechanical services and weather conditions.
- Normal practices used by contractors in installing the insulated enclosure.

Building codes Cold store insulation systems in the USA must meet the requirement of local building codes. These codes relate to the following requirements:

- When foam plastics are used, they must have an acceptable level of flame spread, fuel contribution, and smoke development properties. They must also be covered with facing materials that will guarantee a bond line and the structural stability of the panel.
- When panels are used as exterior wall or interior partitions, they must meet strength requirements for transferring wind loads to steel frames. Generally, they must transfer the wind load to the steel frame without bowing under the load more than 1/240th of the length of the panel. An example of this might be that a panel spanning 6 m vertically between supports should have a bow less than 28 mm when subjected to an approximate 100 kg/m^2 wind load.

Conference of Building Officials During the late 1960s, insurance companies in the USA took an interest in establishing a building code for the construction of refrigerated warehouses which, at that time, was related to general building materials and fell under the auspices of the Conference of Building Officials (CBO).

The CBO established a variety of tests primarily for investigating the fire properties of insulated panels. The best known of these were the 'corner tests', which were undertaken by a variety of testing institutes normally related to the major insurance underwriters. Factory Mutual Research and Underwriters Laboratory—both independently established research facilities—were the two major testing facilities utilized in the USA for the Establishment Fire Testing Certification. Results established by fire tests undertaken by these two major research institutes were accepted by most insurance companies as documentation sufficient to establish the fire properties of various insulated panels.

In contrast to the Lloyds approach, the Factory Mutual and Underwriters Laboratory approach related only to the fire properties of panels and not necessarily to the jointing and construction techniques, nor their long-term insulating value. Whenever general acceptance of building materials was necessary, these fell within other areas of the International Conference and appropriately insulated panels were often compared with other more generally used building materials.

5.17 Current development trends

Industry needs

Considerable developments have occurred over the last 25 years in the design and construction of cold stores, and users have discovered to their cost that insulated structures will deteriorate if the wrong material is chosen.

Generally, if the insulated structure remains intact and free of ice for many years, it usually means that the integrity of the panel joints has remained constant and that the vapour seal has not broken down. It therefore suggests that the bondline between the insulation and the inner and outer skins of metal will remain unbroken as long as it remains dry and free of ice.

Fire retardancy   Although 'flame retardant' grades of insulation do inhibit ignition from a small source, they may burn fiercely if they are involved in a significant conflagration. Consequently, all panels should have an incombustible skin to both sides of the insulation as the minimum of fire protection. Reference should be made to local building regulations before assuming that any particular system will comply.

Potential problems   We have dealt earlier in the chapter with various types of insulation recommended for panel construction. Some cold store users have a preference, normally based on previous experience. However, we should look beyond the need for a particular insulation material and concentrate on the real problems experienced by the operator.

Assuming that the thickness of insulation has been calculated correctly, taking into consideration that this is largely based on experience from relating the capital cost of the actual insulation to the running cost of absorbing the conducted heat by refrigeration, we should then establish that the areas most likely to cause problems within the construction are the vapour seals, and the structural integrity of the panels and doors.

Without doubt, the most common problem experienced over the years is the breakdown of the insulated envelope due to water absorption in the panel which then freezes at low temperature. Large ceiling areas have been known to collapse well within their expected life due to the insulated core being almost solid ice. The weight of the panels has been known to exceed the design weight by as much as five times.

If the insulated core is faced on the external side with steel, then the only area where water vapour can penetrate is the joints between the panels. This suggests that the most important factor in the cold store design is to ensure a positive vapour barrier combined with a structurally sound panel.

The importance of doors should not be neglected. The cold store operator will be forever frustrated if the wrong doors were chosen. It is important to establish the number of door openings per day, as this is a deciding factor in the choice of automatic or manually operated sliding doors.

Doors should be kept closed at all times when not in use, and consideration should be given to the fitting of PVC strip curtains on the interior of the door opening to reduce the level of moisture entering the store while the door is in operation.

To ensure that the cold store user obtains the correct specification for his particular application, he should be aware of all new developments in the industry appertaining to panel design, and should satisfy himself, by exchanging information with manufacturers and contractors, that what is proposed meets with the United Kingdom standard codes of practice.

Structural alterations

When it is necessary to make structural alterations to existing cold stores, consideration should be given to the following points:

- Has the existing system proved satisfactory?
- Can the store be extended or altered without detriment to the stability of the existing structure?
- Can the alterations be carried out with the minimum disturbance to the existing operation?

Consultations with contractors at the early planning stage can prove to be very cost-effective, with advice given on recent developments which could overcome difficulties that may previously have been experienced.

Alternative finishes for cold store panels

Steel finishes   The finishes available for steel-faced panels as produced by the British Steel Corporation are:

- *Colourcoat Pvf 2*   This is used mainly for roofing and cladding applications. The Colourcoat system for external use consists of a single weathering layer on top of the primer and pre-treatments which are applied to the galvanized substrate. The coating is resistant to chemical and solvent attack and has good heat resistance. The good colour retention makes later extensions to the building less conspicuous. Colourcoat Pvf 2 has a life expectancy of over 40 years if the weatherside of the material is properly maintained.
- *Colourcoat Plastisol*   This is more commonly used for panel construction as it can be used on both sides of the panel and on either internal or external insulation systems. Colourcoat Plastisol is a PVC coating on a galvanized substrate and, as with all Colourcoat materials, can be supplied in various colours. It has a similar life expectancy on the weatherside to Pvf 2.
- *Stelvatite*   This differs from Colourcoat in being an organic film laminated to the steel by an adhesive. It has a hard-wearing surface and is normally used on the internal face of the panel. It is more commonly used as a 'Foodsafe' finish when the application is likely to be in contact with fresh meats and poultry.
- *Silicone polyester*   This was developed primarily as a low-cost cladding

and roofing material with a medium-term life for worldwide application. The coating has good resistance to ultraviolet light and heat, but care should be taken in its use near the sea or in hot humid conditions.

- *Architectural polyester* This is a flexible economic cladding material with a medium term life in most non-aggressive environments. As with silicone polyester, if this material is given regular maintenance, its life expectancy can be increased considerably.

Steel colours On externally insulated systems of construction where skin temperatures significantly affect the level of thermal bowing to the insulated panel, consideration should be given to the colours available and the choice of colour that will best reflect the sunlight. It is generally acknowledged that light tones are better for external use, not only for the roof but also for the wall cladding. By using the colours recommended by the steel manufacturer, you will not only prolong the life of the cladding but also improve the effective 'U' value.

Materials suitable for extreme climatic conditions

When designing and insulated enclosure for more extreme climatic conditions, it is well to establish the following:

- Dimensional stability of the insulation if exposed to excessive heat
- Suitability of external cladding for the particular environment
- Necessity for fully galvanized steel fixings and ancillaries.

Generally the methods of construction should not differ from those in the UK as long as adequate protection is given to the components that are exposed to extreme conditions.

Thermographic scan

Once the cold store has been reduced to temperature, it is possible to establish any heat leaks between the panel joints and the panels themselves. An infrared camera is used to detect leaks. The scan should be interpreted by the insulation contractor, preferably under the direction of an independent specialist firm who can obtain the appropriate certification. The scan should be carried out from the interior of the store, which should be free of all racking and at the operating temperature with the doors closed and the fans off. The store should have been at operating temperature for approximately 48 hours before the start of the tests. The panel joints and the panel should be separately scanned and the instrument used should be accurate to $\pm 0.1\,°C$. The scanning technique does not give quantitative values, but effectively indicates zones in which rectification work, particularly at joints, can be observed.

6 Refrigeration plant

H.M. HUNTER

6.1 Introduction

Refrigeration may not have been the most spectacular development in mechanical engineering, but its impact on society is very significant. In the UK the Industrial Revolution transformed the country from a mainly agricultural nation to a manufacturing one and this accentuated the need for preservation and transportation of food. As urban growth developed, it became increasingly difficult to feed the nation from the produce of the traditional rural economy. Country people who could afford it enjoyed fresh meat, and those living by the sea had easy access to newly caught fish. Those in towns, however, were dependent on horse-drawn transport, and preservation by smoking or pickling developed. Taste was often improved by spices, but the demand for highly spiced food diminished in the eighteenth century. The need for food preservation, however, increased.

Many of the early installations of equipment were on board ship, mechanical plant replacing vast quantities of blocks of ice. Both Britain and Australia keenly pursued the idea of refrigerated transport for meat. This is not surprising, as Britain has too little meat, and Australia too much. Furthermore, there was a vast distance between them. The Americas also had surpluses of meat for export, and the advent of mechanical refrigeration brought them great benefit. Hitherto, meat had been largely regarded as a waste product. Because there was not a sufficiently large market for meat in their own relatively lightly populated areas, the producers bred their beasts for wool and hides and scrapped the carcasses. Any price the farmers could obtain was therefore a gain to them instead of the former 'dead loss'.

The preservation properties of ice have been common knowledge for a long time, but it was not until the mid-nineteenth century that developments in mechanical refrigeration started. Initially, simple single low-temperature systems were used to transport perishables from areas of production to those of major population. James Prescott Joule and William Thomson (later Lord Kelvin) really started mechanical cooling by demonstrating that the expansion of gas in a vacuum resulted in reducing temperature.

In general, refrigeration is defined as any process of heat removal. More specifically, it is that branch of science dealing with the process of reducing and maintaining the temperature of a space or material below the temperature of the surroundings.

The rate of which heat must be removed from the refrigerated space or material in order to produce and maintain the desired temperature is called the *heat load*. In most applications this is the sum of the heat that leaks into the refrigerated space through insulated walls, ceiling, and floor, the heat that enters the space through the door openings, and the heat that must be removed from the refrigerated product in order to reduce the temperature of the product to the space or storage conditions. Heat given off by people working in the refrigerated space and by motors, lights, and other electrical equipment also contributes to the load on the refrigerating equipment.

The ability of liquids to absorb high quantities of heat as they vaporize is the basis of the modern mechanical refrigerating system. As refrigerants, vaporizing liquids have a number of advantages over melting ice. The process is more easily controlled, the rate of cooling can be predetermined and the vaporizing temperature of the liquid can be governed by controlling the pressure at which the liquid vaporizes. Moreover, the vapour can be readily collected and condensed back into the liquid state so that the same liquid can be used over

Figure 6.1 Cold air refrigerant apparatus.

Figure 6.2 Piston vapour compressor.

and over again to provide a continuous supply of liquid for vaporization.

The first systems used cold air as a refrigerant (Figure 6.1) At low temperatures, however, this arrangement was not efficient and the vapour compression machines utilizing carbon dioxide or ammonia soon became available (Figure 6.2). Although ammonia was, and is, a very efficient refrigerant, its toxicity and unpleasant smell have prevented its use, or seen its use diminish, in many applications. Neither carbon dioxide nor ammonia was really suited to the smaller refrigerating compressor, and the use of methyl chloride as a refrigerant became widespread. This was the original halocarbon. It worked at low pressures, was readily available, and could be used in a system exposed to copper and brass. Its flammability and toxicity could be tolerated due to the small quantities used in any system. However, for larger systems some other medium was necessary.

This led to the development of more acceptable refrigerants, and so the CFCs appeared. The CFC refrigerants R12, R22, and R502 are now widely used in cold stores, the choice depending on systems design, compressor selection, efficiency, availability, cost, or size of store.

At present there is considerable debate on the use of these CFCs, particularly R12 and its derivatives, as it is widely thought that their escape into the atmosphere causes, or contributes to, breakdown of the ozone layer which protects the Earth from harmful radiation.

A typical system running on the vapour compression principle consists of the following components:

- *An evaporator*, which provides a heat tranfer surface through which heat can pass from the refrigerated space, or product, into the vaporizing refrigerant
- *A compressor*, which removes the vapour from the evaporator, and raises the temperature and pressure of the vapour to a point such that the vapour can be condensed
- *A condenser*, which provides a heat transfer surface through which heat passes from the hot refrigerant vapour to the condensing medium
- *A refrigerant flow control*, which meters the proper amount of refrigerant to the evaporator and reduces the pressure of the liquid entering the evaporator so that the liquid will vaporize in the evaporator at the desired low temperature.

A suction pipe conveys the low-pressure vapour from the evaporator to the compressor, and a hot gas or discharge line delivers the high-pressure, high-temperature vapour from the compressor to the condenser. A liquid line carries the liquid refrigerant from the condenser to the flow control device. Other components and pipes are normally included, but the above describes the main items. More complex systems, refinements and developments of such systems are covered in detail later.

In essence, therefore, refrigeration takes place in a cycle as the refrigerant changes state several times as it passes through the system.

Equipment has of course changed considerably over the years. Electric motors, rather than steam engines, are now almost always used as prime movers. Compressors have greatly increased in speed, and the piston machine is now rapidly being replaced by the rotary compressor (Figure 6.3). The original horizontal machine gave way to the monobloc, where cylinders were cast in one piece. Motionwork was less elaborate, construction was lighter, and fully automatic control was possible. The development of the veebloc compressor, with cylinders arranged in the form of a V, enabled greater capacity to be generated from a smaller machine by improved design and speed.

Early cold store evaporators were simply lengths of pipe fixed to the cold store walls and ceiling, the refrigeration effect being by convection. Similarly, atmospheric condensers were used where water fell by gravity over coils and pipe. Changes took place in the refrigeration of perishables, and brine grids were augmented in many instances by the addition of fans. This was then gradually superseded by the introduction of ducts to distribute the cold air through the storage space. This change in method required corresponding improvements in the design and insulation of the refrigerated environment.

Now, with both evaporators and condensers, it is most common for air to be forced over, and through, coils by electrically driven fans. Alternatively, in the

Figure 6.3 Rotary compressor.

case of condensers, a shell-and-tube vessel may be used being supplied with water cooled in a forced-air cooling tower.

Since the early days of cold storage, eating habits have changed considerably. Nowadays plant is used to pre-chill, blast chill, freeze, and preserve fresh or frozen (Figure 6.4). These processes can involve refrigerating air, water, brine, glycol, or lowering the temperature of metal for contact cooling and freezing (Figure 6.5). Modern standards demand temperature control during processing and so, in many instances, from raw materials through the retail outlet and into the home, refrigeration is used. Developments in transportation have also dramatically affected the food industry with once local products now becoming available worldwide. Whole new markets are being created. This has resulted in the growth not only of long-term cold stores, but also of distribution stores. The emergence of the domination of supermarkets has brought in new ideas in distribution, and multitemperature composite stores are now required.

System, and therefore temperature, control and monitoring has also come a long way. Exacting standards now have to be maintained in the continuous pursuit of quality. Temperature constraints are imposed throughout the whole of the cold chain and any failure to comply can prove to be very expensive to those who do not meet the required conditions. Electronics have taken over from electromechanical devices (Figure 6.6) and current systems can easily cope with controlling and maintaining a whole group of different stores,

Figure 6.4 Refrigeration plant room — reciprocating compressors.

Figure 6.5 Refrigeration plant room — screw compressors.

Figure 6.6 Electronic controls for refrigeration plant.

located anywhere, by the use of dedicated connections or the public telephone system. A temperature rise, or plant failure, in a store in Australia can be immediately identified at the store, or the manager's home, and also at head office in, say, London if need be. the days of operatives manually checking and logging plant conditions are coming to an end.

The cold store of today is a far cry from the early days of refrigeration of the last century.

6.2 Refrigeration systems

Refrigeration systems are categorized by the method in which refrigerant is fed to the evaporator.

Direct expansion

The flow of refrigerant through the tubes of the evaporator (which may be either an air cooler or liquid chiller) is metered by a modulating control valve (usually a thermostatic expansion valve) so that sufficient refrigerant is injected to provide dry and superheated refrigerant vapour at the evaporator outlet (Figure 6.7). This superheated vapour is returned to the compressor.

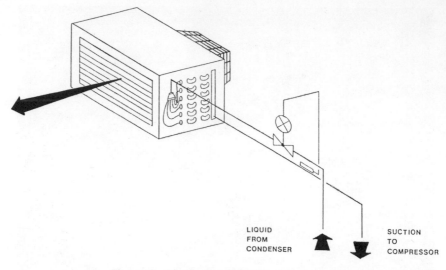

LIQUID
FROM
CONDENSER

SUCTION
TO
COMPRESSOR

Figure 6.7   Direct expansion refrigeration system.

Natural flooded

The evaporator contains a bulk of liquid refrigerant either inside or surrounding the evaporator tubes, in order to wet fully the heat exchange surface. The flooded evaporators are characterized by having a free liquid surface and incorporate a vapour separation device to prevent the entrainment of liquid refrigerant in the suction vapour to the compressor (Figure 6.8). No arrangements within the flooded evaporator are made for superheating the suction vapour.

Pump circulation

Where lengthy refrigerant distribution pipework and/or specialized process evaporators (for example, plate freezers) are necessary, liquid refrigerant pump circulation systems are used. These incorporate a liquid reservoir, generally termed surge drum or suction separator, and refrigerant liquid pump to deliver liquid quantities at least double the quantity evaporated (and sometimes as much as 15 times the quantity evaporated) in order to maximize the usage of evaporator heat exchange surface and overcome pressure losses (Figure 6.9). Both vapour and liquid are returned to the liquid reservoir where the surplus liquid is separated for recirculation to the evaporators and the disentrained vapour is drawn into the compressors.

6.3   Power comparisons and running costs

All the systems described may use either single-stage compression for chill temperatures (0°C) or compound compression for cold storage ($-30°C$).

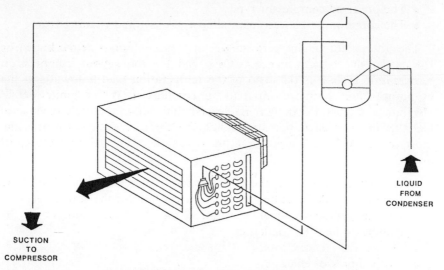

Figure 6.8 Natural flooded refrigeration system.

A number of hybrid systems have been developed, particularly with the advent of electronically-motivated control valves and control circuits, with the object of combining the advantages of the different refrigeration systems.

The decision on either single-stage compression or two-stage compression is dependent upon:

- The choice of refrigerant
- The design, evaporating and condensing temperatures

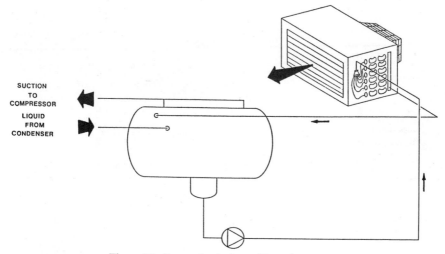

Figure 6.9 Pump circulation refrigeration system.

- The preferred compressor type
- The compromise between capital cost of plant and running costs

The criterion for judging performance of a refrigeration system is known as the coefficient of performance (COP), and for the vapour compression refrigeration cycle it is the ratio of the refrigeration load (kilowatts at the evaporator) and the power input to the compressor. Within limitations of economical factors, the design engineer should obtain the highest value of COP i.e for a given refrigeration load, the engineer designs towards the most economical work input. Factors which effect the value of the COP are:

- The temperatures of the low-temperature region containing the evaporator and the temperature of the coolant for the condenser
- The temperature difference required for the heat transfers in the evaporator and the condenser
- The system pressure losses
- The compressor efficiency
- The refrigerant

The typical range of values for COP for chill storage and cold storage/freezer applications are given in Table 6.1.

Table 6.1 Typical values of COP.

	Refrigerant	COP
CHILL STORAGE	R22	3.4–4.0
	R717	3.8–4.6
	R12	3.6–4.3
	R502	3.2–3.9
COLD STORAGE		
Single stage compression	R22	1.5–1.8
	R717	N/A
	R12	1.4–1.7
	R502	1.6–1.9
Compound compression with intercooling	R22	1.7–2.0
	R717	1.9–2.2
	R12	1.6–1.9
	R502	1.8–2.1

	Evaporating temp.	Condensing temp.	Suction discharge pipe pressure losses
Chill storage	$-6°C$	35°C, ambient 26°C	1°C
Cold storage	$-36°C$	35°C, ambient 26°C	1°C

6.4   Use of waste heat

The heat extracted from the cold spaces, plus the heat energy of the work done by the compressors, is ultimately rejected through the condensers. There are three phases of this heat rejection:

- *De-superheating* of the hot vapour, typically from 80/120°C for piston compressors and 65/80°C for screw compressors, to the condensing temperature. This de-superheating phase represents between 10 and 20% of the total heat rejection.
- *Condensation* or latent heat change at the condensing temperature, (generally between 30 and 40°C in temperate climates, depending upon ambient conditions). This latent heat change represents between 80 and 90% of the total heat rejection.
- *Sub-cooling* of the condensed liquid yields only 2–4% of the total heat rejection and for purposes of waste heat recovery in this context may be ignored.

Typical applications for waste heat recovery are:

- Glycol solution heating for circulation through underfloor grids in the cold stores (Figure 6.10)
- Water heating for washing, cleaning, etc (Figure 6.11)
- Background heating for workshops, carton, or dry stores (Figure 6.12)

The attainable temperatures for the waste heat fluid are not high since the bulk of the heat rejected is at the refrigerant condensing temperature. For

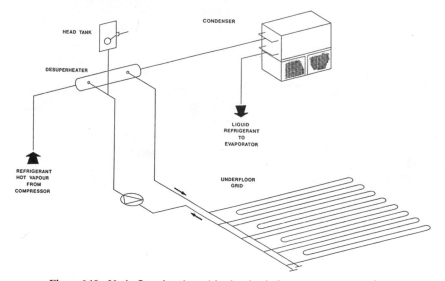

Figure 6.10   Underfloor heating with glycol solution to recover waste heat.

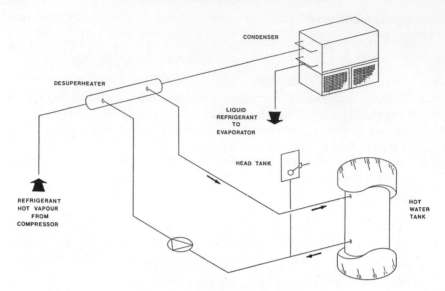

Figure 6.11 Water heating to recover waste heat.

example, with a condensing temperature of 35°C and a 5°C temperature differential for the waste heat exchanger, then water heating is only possible to 30°C to recover this bulk heat. Higher water temperatures from waste heat recovery to 50°C may be obtained, but by using the smaller de-superheating phase of heat rejection.

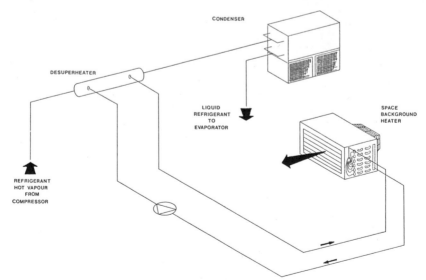

Figure 6.12 Background heating to recover waste heat.

In appraising any heat recovery system the designer needs to recognize that the refrigeration plant may not work at its design rating for much of its life and consequently the values of heat rejection will be lower. In particular it should be recognized that there is usually less demand for refrigeration during the winter months and, consequently, less heat available for recovery. Thus the heat recovery scheme may need to be supplemented by alternative heat sources.

It is quite feasible to recover all of the heat rejection and obtain higher temperatures, up to 80/90°C, by using a high-temperature heat pump. However, in many countries this application seems unnecessary as prime fuel costs are not yet prohibitive.

6.5 Defrosting methods

Four principal methods are employed for air cooler defrosting:

- *Electric defrost* with heaters arranged in the air cooling coil is the most effective form of electric defrost heating and is best suited for use with smaller ceiling-mounted air coolers when the heating power load is not excessive. It can be applied to larger coils provided due account is taken of convection losses. However, the system can be expensive to run especially if peak rate electricity tariffs cannot be avoided (Figure 6.13).
- In *hot gas defrost*, the heat which would normally be rejected by the condenser is diverted to the air coolers. The air cooler being defrosted then acts as an auxiliary condenser. Hot gas defrosting is generally considered to be more efficient than electric defrosting although coil

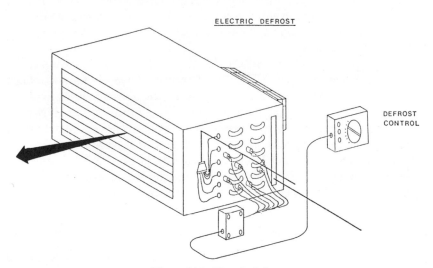

ELECTRIC DEFROST

DEFROST CONTROL

Figure 6.13 Electric defrost.

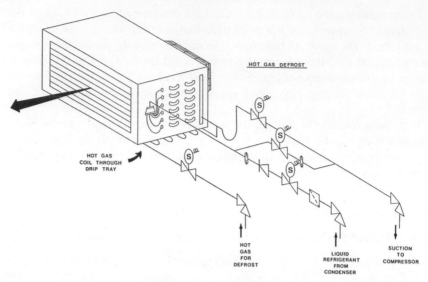

Figure 6.14 Hot gas defrost.

block temperatures achieved during defrosting are lower. Hot gas defrost is the most economic system to operate since the defrosting cycle on one cooler is concurrent with the refrigeration cycle on other coolers (Figure 6.14).

- During *reverse cycle defrost*, the roles of condenser and evaporators are reversed by changeover valves. All air coolers are defrosted simultaneously. This system is generally less costly to install than hot gas defrost but results in a greater rise in cold store air temperature during the defrost sequence and since the compressor and condenser must be operating to effect a defrost, running costs will exceed those achieved with hot gas defrost (Figure 6.15).

- With adequate supply of water at 10°C, *water defrost* can be very effective but its application is generally limited to rooms above −4°C and for blast freezers where the temperature can be raised above freezing without serious inconvenience in the event of a valve malfunction. The air coolers need to incorporate a water sparge arrangement with adequately sized collecting tray and drains. The arrangement of the water headers and pipes within the cooled space is obviously critical, as they must be completely drained after each defrost. Running costs are limited to pump power input and water supply charges (Figure 6.16).

6.6 Monitoring and controls

Even after the introduction of high-speed reciprocating compressors, control systems were still usually very simple. Sophisticated control systems were either very expensive or unreliable. As labour was still relatively inexpensive it

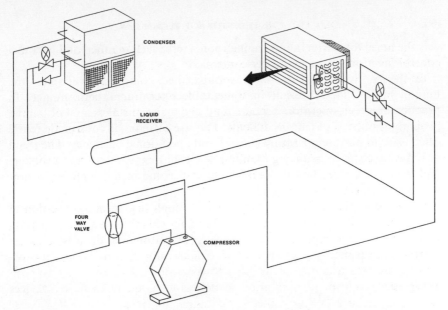

Figure 6.15 Reverse cycle defrost.

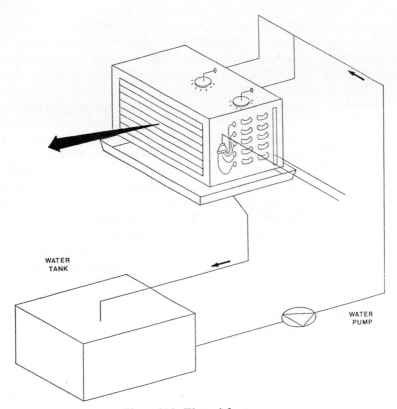

Figure 6.16 Water defrost.

was the norm for refrigeration engine rooms to be well manned and for most control functions to be manually controlled.

In theory, the use of the most sophisticated computer available—the human brain—should result in unbeatable operational performance. In practice, however, operators become tired and make mistakes, and obtaining properly trained staff can be difficult. The use of manual control therefore often leads to poor temperature control and operational efficiency. The trend of reducing costs by reducing manning levels has exacerbated the problem, and the results of inefficient operation have become increasingly important with rising energy costs.

Present-day plants therefore incorporate a high degree of automation in order to minimize manual intervention. Thus, for example, manual regulating valves have been replaced by a variety of automatic devices such as thermostatic expansion valves (now electronic), pilot-operated floats, motorized valves, etc.; treatment of cooling tower water is usually achieved with automatic dosing plants; initiation of defrosting is likely to be by devices sensing ice build-up or air cooler air-side pressure drop.

Automation of this type enables manning levels to be sucessfully reduced and more efficient and repeatable operation to be achieved. However, the revolution in microprocessor technology has resulted in ever more sophisticated control systems becoming available at prices which are low in relation to the achievable costs savings. This, allied to the parallel development of plant monitoring, has opened the way for completely unmanned refrigeration systems.

The first priority of a monitoring system is to make sure that the product reaches the customer in the best possible condition. In case of frozen products, this means checking that the temperature has not risen above a specified limit. With chill products, especially fruit, the temperature limitations are much more demanding. Too low a temperature leading to freezing may be more serious than too high a temperature. Fruit, particularly apples and bananas, require control of the atmosphere (for example a reduction in oxygen level) to prolong storage life and control ripening. (see Chapter 3).

It is desirable to monitor the temperature of a store without having to go inside. If a simple temperature indicator just outside the store is all that is required, then a dial thermometer connected by capillary tubing to a sensing bulb in the store may be fitted.

If, on the other hand, there are a number of stores to be monitored, or distance is a problem, or a permanent record of the storage conditions is required together with an alarm system, then an electrical system might be chosen. The temperature is usually sensed as the change in resistance of a very small coil of platinum wire. Platinum is chosen because it gives very repeatable results; the coil is so small that the cost of the platinum is negligible. A large number of manufacturers supply sensors to the PT100 standard characteristic which are interchangeable.

Sensors from all over the store are connected by ordinary copper cable to an alarm and logging equipment in a convenient location.

The second priority of a monitoring system is to check on the refrigeration plant itself. It pays to keep refrigeration plant operating at peak efficiency. The refrigeration plant for a medium-sized cold store containing say 500 tonnes of produce will use about 20 000 units of electricity per month. A plant operating 20% below its design performance will probably perform adequately enough to keep the product down to temperature under most conditions, but the running costs will have risen considerably.

In days gone by, it was normal to employ a full-time operator to look after and monitor a plant. Nowadays this is a luxury we can no longer afford. Electronic equipment is now available, such as APV Hall's Fridgewatch, which will monitor both the refrigeration plant and the stored product automatically. Besides producing a display of the current values it produces a permanent printed log at regular intervals. Any parameters outside normal operating limits will create an alarm. If required, when such an alarm occurs, Fridgewatch can automatically dial a call over the public telephone system to a service organization or an engineer 'on call' to give warning of a fault before the problem has begun to affect the storage condition of the product.

The monitoring equipment can also control the refrigeration plant, starting and stopping compressors and loading them up to control the temperature of the store within the required limits. It can also control the defrosting of the coolers, determining when a defrost has become necessary and carrying out the sequence of actions necessary to perform the defrost. No installation of any size these days would be complete without its monitoring system.

6.7 Maintenance

A modern cold store refrigeration plant is a highly sophisticated thermo-dynamic process system. Whilst modern equipment is very reliable, proper maintenance is essential if efficient performance is to be achieved and maintained. Maintenance falls into two broad categories:

- Planned or preventive maintenance
- Crisis maintenance

Too often in the past, preventive maintenance was considered an expensive luxury and equipment was only looked at when it broke down. This approach leads to unnecessary catastrophic plant failures and expensive interruptions in production. Thankfully plant operators are now more enlightened, and greater emphasis is being given to preventive maintenance.

Traditionally this was a question of a disciplined approach to checking and repairing (or replacing) plant items on a time interval basis before their performance fell below acceptance levels. Whilst this is very cost-effective, methods have gradually been introduced to monitor or predict the loss of performance of a plant item. These methods are generally known as *health monitoring*. Some methods, such as vibration analysis, have been used for a number of years to monitor bearing wear in compressors and pumps; however, it is the advent of modern monitoring systems which has enabled on-line performance assessment to become a reality.

The main advantage of the health monitoring technique is that equipment need only be serviced when its performance has fallen below preset levels. When items are serviced on a time interval basis, quite often equipment is taken out of service and stripped down unnecessarily.

As mentioned above, modern monitoring systems can continuously assess plant performance criteria such as COP. This information can be 'trended' and retrospective logs produced when a fault is detected. In this way detailed information on plant performance can be presented to plant operators, engineering staff, and management. It can also be passed via modem links to a service organizations and form part of a service or maintenance agreement.

The present-day plant engineer can therefore operate with the minimum of staff but achieve much higher standards of maintenance and operating efficiency than ever before.

6.8   Current development trends

Emphasis on rotary compressors

Just as the veebloc compressors superseded the slow-speed monoblocs, modern screw compressors are rapidly replacing reciprocating compressors. This trend began with larger-capacity machines but now extends down to the smallest industrial sizes.

Screw compressors have existed for some 40 years, but it is largely the increased emphasis on operating costs and efficiency which have led to their increased use today. The ability to have infinitely variable capacity reduction enables very close control to be achieved. This can often result in operation at a higher suction pressure and hence better coefficient of performance.

Compared with reciprocating compressors, screw machines have much fewer moving parts; this contributes to greater reliability. It is not at all uncommon for screws to operate for 25 000 hours without major overhauls, something that would not be achievable with modern reciprocating machines. The reliability is very good, and downtime for maintenance, when required, is also much reduced. This is especially true of the newer single screw compressors. With these machines complete overhauls can be undertaken without disturbing either the drive motor or any pipework.

Variable-speed drive motors

Varying the speed of the motor driving a compressor or pump is an elegant method of controlling the duty performed.

Motors using direct current have been used for a long time, for example, for railway traction. They operate over a very wide range of speeds and torques but the maintenance requirements of the carbon brush gear and commutators, which are an essential part of the design, have discouraged their industrial applications.

Unfortunately, the induction motor which is extremely common, efficient, simple and reliable, operates only at a fixed speed determined by the supply frequency and the number of poles. An induction motor can be wound in such a way that the number of poles can be varied by altering the way the windings are connected. This will make it run at one of several speeds. There must be an even number of poles (N and S) so the selection of speeds that can be achieved is rather limited. Where this limitation does not matter, for example for condenser fans, this is the cheapest and most common method of achieving variable speed.

With the spectacular developments in the semiconductor industry it is now quite practical to consider converting the normal 50 (or 60) Hz supply into a supply where the frequency can be varied continuously from 0 to well above 50 Hz. A normal induction motor running on such a supply becomes a true variable-speed motor. It is necessary to vary the supply voltage at the same time to avoid the motor magnetic circuit saturating at low speed, but the electronics can be arranged to do this. It will also be necessary to incorporate some form of high-speed overcurrent protection. A semiconductor device can be damaged in a shorter time than a fuse takes to operate.

Such a variable-speed drive has a number of side benefits. By 'ramping up' the speed of the motor from standstill, the surges and wear and tear associated with the conventional forms of motor starting are avoided. This is often referred to as a 'soft start'. By running at a frequency in excess of the normal 50 Hz, extra output can be obtained from the compressor, etc. being driven. Drives are widely available with ratings from a fraction of a kW to 1 MW.

Although the cost of semiconductor variable-speed drives has fallen over the years, they are still very expensive compared with fixed-speed drives. Unless there is something very special about the application they do not compete with other methods of controlling the output of the plant such as the modulating slide valve of a screw compressor or the cylinder unloading of a reciprocating compressor.

Electronics

The explosive growth of microprocessors in the late 1970s and early 1980s has now slowed down. Microprocessor-based products have begun to mature and are expected to be as well designed as any other product.

It is true that there has been a considerable fall in the price of office computers such as the IBM PC and its clones, which has radically changed the price of computing power in this sector. However, the packaging and internal structures of these small computers do not lend themselves to the industrial environment, and their main application has been the provision of management information away from the plant-room environment.

Pressure and temperature transducers are now highly developed, and it seems unlikely that any major new techniques will emerge in the next few years.

There is still a need for a really reliable, easy-to-use, and cheap frost detector. Some new developments have appeared in this area lately and it will be interesting to follow their progress. There is also a need for a reliable easy-to-use and cheap Freon detector. The growing concern about the release of CFC into the atmosphere will increase the emphasis on this development. There seems little doubt that the solution to both these problems will be electronic in nature.

System diagnostics

Present systems have automated the collection and recording of data on the operation of the plant, but they rely on a human being to interpret the information and draw any conclusions.

In theory, it should be a simple matter to automate this process as well. However, this is easier said than done. How often is the fault-finding section in an instruction book consulted? Perhaps this is because the faults that Nature creates are a great deal more subtle than those contemplated by the writer of the instruction book.

However, there is unmistakable evidence that there is a diminishing number of engineers with the time, skill, and inclination to analyse a log sheet. And with such a need, a product will surely come.

Today's equipment will already calculate continuously the COP of a plant (the cooling produced per unit of input energy). This can be regarded, rather like fuel consumption per distance travelled by a vehicle, as an indicator of the health of the mechanism. But this is just a start; perhaps in the year 2000 equipment will give a description of the fault, a list of parts and special tools needed, together with a printout of the necessary drawings and diagrams!

6.9 Safety

British Standard 4434, Specification for Requirements for Refrigeration Safety was originally published in 1969, prepared under the authority of the Refrigeration Industry Standards Committee. It was modelled on a draft ISO Recommendation on *Refrigeration Safety, R1662*. The American Standard ASAB 9.1: 1964 was also used for reference.

In 1979, BS 4434 was revised to include all changes up to that date. The revision was prepared under the direction of the Refrigeration Heating and Air Conditioning Standards Committee as BS 4434:1980. The standard continued to include relevant technical data obtained from the International Standard Recommendation ISO/R1662. For several reasons Britain had not been able to accept ISO/R1662 when it was formally issued, mainly due to a lack of clarity between mandatory and advisory clauses.

BS 4434:1980 specifies requirements for the safety of refrigerating systems and ancillary equipment and is applicable to new refrigerating systems, to extensions and modifications of existing systems and to used systems on being reinstalled and operated on another site. It also applies in the case of the conversion of a system for use with another refrigerant e.g. R40 to R12, or ammonia (R717) to R22.

In essence the standard defines five categories of occupancy, six types of cooling system, and three refrigerant groupings. It then specifies certain restrictions applicable to various combinations of occupancy, cooling system, and refrigerant group. The standard also covers working and test pressure requirements and protection against excessive pressure. BS 4434 is at present being revised, to ensure that it is in line with latest technology and practice and also presents the data in a more logical sequence.

Almost in parallel with this revision, the ISO Safety Recommendation R1662 is being revised with the intention of producing it as a standard rather than a recommendation. This standard concerning the safety of refrigerating systems takes into account regulations already in force, or existing in draft form, in a number of countries. The provisions represent minimum requirements for the design, construction, installation, and operation of a refrigerating plant; however, in particular cases, more severe requirements may be necessary. Where national regulations are in force, full account should be taken of them.

Many countries have very comprehensive national safety codes, a number of which were submitted to the International Committee for consideration in preparing the revised standard. Discussion has taken place over several years and participating countries included Czechoslovakia, France, German Federal Republic, Japan, Russia, Sweden, Switzerland, UK, USA and Yugoslavia. A draft document was submitted to the ISO Central Secretariat in April 1987. At some future date it is anticipated that this draft revision will be issued to ISO member countries for acceptance or rejection.

In around 1979, the Institute of Refrigeration decided that the requirements for refrigeration safety given in BS 4434 should be amplified and extended into the form of a specific Code of Practice for establishing mechanical integrity in the larger vapour compression refrigerating systems using ammonia as the refrigerant. The Code is intended to complement and reinforce details in BS 4434, and nothing in the Code is intended to conflict with British Standards. The code is published in two parts: Part I, *Design and construction of systems using ammonia as the refrigerant*, and Part 2, *Commissioning, inspection and maintenance*.

In 1984 the Institute of Refrigeration published a further Code of Practice to define minimum requirements for safety in the design, construction and installation of vapour compression refrigerating systems using R11, R12, R22, or R502 as the refrigerant. Again the document is intended to complement and reinforce details in BS 4434.

In view of the many organizations represented on the committees involved in compiling the above documents, it is widely accepted that when followed they represent good practice within the industry.

A particular aspect of safety which will almost certainly have a considerable impact on the refrigeration industry in future years is the effect of some refrigerants on the atmospheric ozone layer. Regulations already exist in some countries to limit the production and use of gases which are believed to affect the ozone layer (CFCs 11, 12, 113, 114, 115). This will lead to alternative refrigerants being used and the development of new refrigerants. Regulations on the containment and collection of these refrigerants are likely in future. The EEC has issued a code of good practice for the reduction of emissions of CFCs R11 and R12 in refrigeration and air-conditioning applications (Report EUR 9509EN). These guidelines are likely to be more widely applied as time passes.

Safety in refrigeration is a moving target, constantly changing as the technology changes and new developments are introduced. Engineers must remain vigilant to ensure that adequate safety standards are maintained, in order to safeguard the public and to preserve the extremely good safety record achieved by the industry over many years, even with the vast use of refrigeration equipment in almost all walks of life.

6.10 Quality of refrigeration plant

To achieve quality or a level of excellence as perceived by its purchaser a refrigeration plant must:

- Achieve the specification requirements
- Perform in a safe condition

These two aspects are inseparably combined when assessing equipment in terms of quality. To achieve this goal requires input from two major areas of influence—technical and safety.

Technical assessment of a refrigeration plant with regard to its required duty includes consideration of design in terms of thermodynamics and strength of components at service conditions, leading to input on materials, corrosion, vibration, noise, assembly, installation, testing, commissioning, etc. Inevitably all these factors will be influenced by national standards, codes of practice, advisory literature, and indeed client specifications applicable to the plant in parts or as a whole.

All the technical aspects will be influenced by safety legislation and Health and Safety Executive advisory data, whether implicit in the purchaser's specification or not. In the UK, for example, designers must consider the

Factories Act, the Health and Safety at Work Act, COSHH regulations, noise regulations, electricity regulations, and pressure testing regulations.

To achieve cohesion of such a multitude of influences requires all the supplier's operating routines to be embodied in a planned and logical control system referred to as a quality assurance system.

Operating routines include contract review, design control, document control, purchasing, process control, inspection, and test and recording. By processing each contract or order through this established system it is possible to achieve the specification and safety requirements first time every time, improving both profitability and company status.

Until quite recently, much refrigeration plant was specified solely by its required duty, the specifier relying upon the supplier's name, standing in the industry, experience, and integrity to produce a functional and safe plant. This situation has been changed by:

- Escalating costs of capital equipment
- Escalating labour costs
- Higher awareness of energy conservation requirements
- Lower pay-back periods on capital investment
- Legislation, e.g. Health and Safety at Work Act and COSHH regulations
- Input from Codes of Practice and British Standards, e.g. BS 4434
- Input from insurers regarding stored products, loss of process, and personal damage
- High financial risk from plant failure in industries such as petrochemical, offshore, and cold storage
- The activities of some contractors who pay little heed to quality and safety.

7 Electrical installations

F.W. CARR

7.1 Introduction

Many may think that the electrical installation required for a cold store is basically the same as that for most other industrial sites. Certainly, the wiring and safety aspects have to comply with the same regulations, and cold stores have to be lit to the same ambient level. But there is one important difference and that is the temperature: materials perform differently at $-25°C$ from the way they do at normal temperatures. Certain plastics may crack, standard lights are less efficient, and without care the cold store could freeze the ground on which it is standing and cause foundation problems.

As with any installation, there is a wide exposure of personnel in the working environment to the electrical services, each person making his own demand on power, lighting and emergency systems. In the temperature-controlled storage industry, a high percentage of overheads is represented by electrical consumption. Sometimes the energy bill can represent 40% or more of the cost of running a cold store, even without the capital costs of installing electrical services.

The cold store industry cannot afford to economize on the electrical installation, which is a small cost when compared with the total value of frozen food products that will eventually be stored. But this does not mean that the costs cannot be restrained, and different ways in which energy costs can be saved by the use of prudent energy management techniques will be considered later.

The ever-increasing developments in electronic control systems, computers and communication systems have given cold store managers and engineers some of the tools that they can use to increase efficiency and lower cost in their cold store operations. But these advanced systems bring with them their own problems and make their own demands on the electrical installation. Clean earths and dedicated circuits for example are needed to help protect the delicate electronics from the effects of high-voltage mains which are used in other parts of the system; voltage regulators and uninterruptible power supplies become essentials, not luxuries.

Electrical systems play a crucial part in the safety of a site. Functions such as emergency lighting and fire alarm systems are important on any site, particularly so with cold stores where there are the added dangers of workers being trapped at low temperature or at risk from toxic gases. The electrical system plays a vital role not only in protecting the plant but also in saving human life.

Although the increased sophistication in electrical and electronic apparatus is creating tremendous benefits in safety and cost-saving, its inherent complexity demands the highest standards of engineering and installation. Many professional and statutory bodies have requirements that must be adhered to within a cold store.

All in all, a well-designed and engineered electrical services installation will provide the cold store operator with a safe, pleasant and economic working environment in which to store product and employ staff. But a badly designed installation, which does not recognize the individual needs and mode of operation of the cold store operator, can induce higher than necessary energy costs, poor standards of safety and labour and personnel problems.

This chapter will discuss the general considerations of electricity supply and distribution—including tariff structures—and how to obtain the best from them. It will continue with a general overview of how a cold store works, and then go on to specific problems of lighting, safety, security, stand-by power, computer installations and energy saving.

7.2 Electricity: the starting point

Generation

Electricity is generated as alternating current (AC) at 50 Hz or 60 Hz because it is easier to transform from high to low voltages with AC than it is with direct current (DC).

The two most common types of generating plants are coal and oil stations, where the fuel is burned to produce steam that turns a turbine on an electrical generator. There is also a growing number of nuclear stations, in which the heat from the atomic reaction serves the same purpose as burning fuel. Alternative power sources are water-powered stations, wind-powered stations and solar-, wave- and tidal-powered schemes.

Transmission

In the UK, the main grid has the electricity transmitted at voltages as high as 750 kV. This is transformed via intermediate voltages to local substations at 11 kV. It is then further transformed to 415 V three-phase or 240 V single-phase for most consumer premises, although many industrial commercial premises, including cold store facilities, take their power at 11 kV.

It may be considered, if the anticipated demand is high enough, for a factory to have its own on-site substation or series of substations. Many factories are installing generating sets, not only to provide a back-up during power cuts but also to generate their own electricity during Area Board peak tariff rate times and to avoid exceeding agreed maximum demand levels.

Within a site, power is distributed on a number of ring mains or radial circuits. These circuits are independent electrical feeds each of which powers a certain area of the site or a particular type of appliance. For example, the cooling equipment may be on a different circuit to the lighting. When allocating appliances to each circuit, essential and nonessential supplies have to be considered. At peak times, for example, the user may wish to turn off, manually or automatically, all nonessential equipment in order to save money. If there is a power cut, the back-up generating set may not be able to handle the complete capacity of the plant, so power demands have to be given priority.

An example of an essential circuit would be one which powers the fire alarms, emergency lighting and other safety equipment (although much of this may be on its own battery system). Computer equipment is also usually on an essential service, with an uninterruptible power supply to maintain power while the generating set is being started up.

Tariffs

The charges for electricity can vary greatly between different times of day; tariffs may be based on the average consumption at the various times. At peak times, the charges are at their highest and at off-peak times, such as the middle of the night, the charges are at the lowest. It is always worth consulting the supply authority for details of its tariff systems and advice on how to make the best of low-cost off-peak energy.

Many cold stores turn their refrigeration units on full for a short time before the end of the off-peak night time period, to cool the stores to the lower temperature limit of the store during the night. This means that the refrigeration system can operate at a lower energy consumption during the day—when tariffs are higher—while the store warms back to the upper temperature limit.

Metering

The supply authority will provide the necessary metering equipment for all tariff rates. In addition, the consumer may wish to install his own industrial meters to check kilowatt hours consumed, maximum live demand, or relative consumptions of different site operations, for example.

7.3 Control systems: computerized refrigeration

The general trend in control systems for industry is towards greater computerization, and refrigerated stores are no exception. Control systems can be divided into three classes:

- Protection equipment—for safety
- Operating controls—such as thermostats
- Controls that monitor performance.

Protection devices are not only for the safety of personnel but also protect the refrigeration equipment, much of which may be operating at high pressures. The compressor, for example, is the heart of the refrigeration system. This will have protection devices such as high pressure cut-outs, high temperature cut-outs and oil pressure switches. Some of these machines are self-lubricating and so may have to run for some time before the oil pressure reaches the required level; if the pressure fails to reach this level, an oil pressure switch is needed to stop the compressor.

Operating controls are used to maintain a desired temperature in the cold store, or pressure in the compressor plant.

Monitoring controls often take the form of chart recorders; they may measure temperature at key positions within the controlled areas, or floor and ambient temperatures to check the stability of the insulated areas under working load conditions.

Temperature recording and control

There are three basic types of temperature recorder:

- Capillary probe
- Period chart recorder
- Continuous chart recorder

The simplest temperature recorder is the capillary probe, which fits through the cold store wall and operates a visual indicator; this is nonelectrical and nonmechanical, rather like a normal thermometer.

A chamber of $14\,000\text{m}^3$ (0.5 million cubic feet) typically has two probes in the store. In critical applications, it is sometimes necessary to have probes *inside* the product being stored; after all, it may be the temperature of the goods which is the main criterion. The number of probes is often decided by the owner of the goods stored. A number of hand-held instruments with probes are available and the owner may use these also to do spot checks around the store. There are a number of digital recorders on the market, whereby a probe can be placed in the chamber linked to a digital read-out. These are suitable only when historical records are not required.

Period chart recorders may be simple temperature recorders that store information either as a seven-day or four-week circular chart that gives an overview of the chart period. A continuous chart recorder produces much longer records, although the visible section may be quite short. Temperature chart recorders have changed dramatically in recent years as more and more manufacturers take full advantage of microprocessor technology which enables analytical data to be down-loaded into very sophisticated computer control systems. These allow the different needs of cold store operators—for different degrees of accuracy, numbers of probes, high and low alarm values, product recovery trends—to be met.

Integrated into all control systems are various alarm units and vapour detection systems. These would normally be controlled from a central panel that can be checked, altered and monitored by a refrigeration engineer. Before the days of microelectronics, electromechanical relays, arranged to perform the logic functions, would have handled the process, but with the advent of programmable controllers all this is changing.

The change is being tackled in two ways: the first is to use equipment which has already been developed, such as a Sprecher + Schuh 530, but programming it to suit the control system requirements; the second is the development of dedicated systems purely for the control of refrigeration stores. One such model is the Telstar from Star Refrigeration.

The Telstar microcomputer was developed to control one or two compressors in a simple refrigeration system, but it has already been used to control systems with up to six small screw compressors. The system is also capable of expansion by using a number of Telstar controllers that can interact with a programmable controller and report to a common transmitter.

The Fridgewatch M system from APV Hall International was developed from the Sentralink microprocessor system for 24-hour monitoring of refrigeration plants. Transducers monitor the running conditions on each compressor, and the transducers are connected to input cards which provide pressure cut-out signals directly from the transducers. The system also receives status information from contacts in the starter panel, and can calculate and display refrigerant superheat, hours run, number of starts and approximate energy consumption.

Telstar, Fridgewatch and similar proprietary systems can be connected to standard computer peripherals, such as printers, and can be linked to a modem to send information over telephone lines.

The advantage of microelectronic systems over the old relay method is evident when it comes to alteration or expansion. With relays, these changes involve a significant hardware problem, but with microprocessor based products they can be achieved by altering the software. The new systems can handle analogue signals via in-built analogue-to-digital converters.

The problem with relay systems mean that many users are finding it more economical to use programmable controllers, particularly as solid-state electronics are more reliable.

Programmable controllers

Energy costs are a substantial part of running refrigeration equipment for a cold store and it is essential that the temperatures of the stores are closely

regulated, particularly at the low end of the range. If the desired temperature is
$-29°C$, the cost of running it even a degree or two lower is significantly
greater. By using temperature transducers via an analogue-to-digital converter in the programmable controller, highly accurate temperature regulation
can be achieved.

Analogue devices can be installed to measure build-up of ice in the coolers
and provide auto-defrosting. The programmable controller can measure this
and carry out the defrosting at the most convenient and economical time. One
potential problem with defrosting is that if the heater on or within the drain
taking the defrosted water away from the cooler fails, then the drain will
become blocked with ice and water may escape into the store. This water may
then re-freeze and create a major problem. With a programmable controller,
current-sensitive devices can be used to monitor the current on the drain
heater and to sound an alarm if the heater fails.

A programmable controller can be used to provide maximum demand
control by shutting off equipment in a predetermined order when a pre-set
maximum demand level is approached. Similarly, it can run the plant to give
the best use of electricity board tariffs. A progammable controller can provide
event recording, either on a printer for hardcopy records or onto a screen for
instant monitoring. The event recorder provides a timed message which gives
equipment starting and stopping times, values of temperature and pressure
and alarm messages. The alarm message feature is particularly useful. If an
alarm message is received, the previous read outs will provide a state-of-the-
plant report at the time the fault occurred and this can help in finding out
what went wrong.

The controller can be used to ensure that running times of the compressors
are approximately equal to allow their essential checks and oil changes to be
done in one maintenance period, creating less of a disruption in the running of
the plant.

One problem with some types of plant is that of a high level surge within the
surge drum which can produce a potentially dangerous condition. When the
controller receives such a high level signal, it can take appropriate action, such
as shutting down all or part of the plant and giving an alarm signal.

The controller can also be used for data logging to provide, for example, a
four-hour log showing minimum/maximum values of temperature and pressure,
the number of alarms during the period, the electricity consumed and average
values of all of these; at the end of the day, these could be collated as a 24-hour
log.

The future for the cold store industry is clearly towards more and more
microprocessor-controlled equipment. The greatest trend is towards remote,
centralized control, where a number of plants are monitored via telephone
lines from a central location. This could mean that it would no longer be
necessary to have a skilled engineer on each site. Instead, a maintenance
contractor could simply visit the sites and send reports to the centre using a
modem on the telephone line.

Figure 7.1 Composite control and distribution panel at a large distribution depot. Fed by two transformers, the incoming circuit breakers are situated either side of the control desk, with an off-load bus coupler between them. The copper bus bars are top mounted in the ventilated enclosure, and the heavy plant and distribution loads are split to both sides of the control centre thus balancing the transformer loading. All cables enter and exit via floor ducts, the covers of which can be seen.

There is also a move towards greater capabilities of the programmable controller such that the gas detection equipment, chart recording, alarm systems, control gear for the drives and other ancillary equipment of the plant may all be incorporated in a composite control panel located away from the plant room (Figures 7.1 and 7.2).

Computers for office and plant

One of the biggest problems in running any type of computer system in a factory is the closeness of the data cables (carrying computer information) to the main power cables. High voltage cables produce electromagnetic and radiofrequency interference which can corrupt and destroy data on a computer cable. This means that all the cables should have special shielding around them that cuts out the spurious signals.

Figure 7.2 Detail of the control desk shown in Figure 7.1. A programmable logic controller is used for the central refrigeration plant control and overall monitoring. A VDU gives a graphic display of temperatures and pressures throughout the system, a printer records events and a keyboard gives the Site Engineer access to the system. An electronic gas detection system can be seen behind the printer and, above it, two strip chart temperature recorders. The whole control package is grouped to provide the engineer with convenient display and access to information.

Heavy electrical equipment being turned on or off can cause sudden jumps or falls of power on the mains. These are called 'spikes'. Spikes can be transferred from the mains cable to the computer and this too can destroy data on the computer or even cause it to crash. The computer should be put on a clean-earth system where its mains are separate from the rest of the plant. And, as will be mentioned later, if an uninterruptible power supply is fitted this will protect the computer from the spikes. There are also mains filters on the market which can do this job.

The computer's central processing unit in particular should be protected by using one of the above systems. Some memory circuits can be protected from loss of power with their own built-in batteries and it is sometimes safer to put some, or all, of the software used onto chips called EPROMs which do not lose memory when the power is off.

The problem with these is that to change the software means reprogramming the EPROM which normally has to be done outside the machine with special equipment.

It is not only the computers in the office that need protection. A programmable controller too contains microprocessors and other sensitive electronics. It may be used to control all the refrigeration and motor facilities as well as acting as an energy manager to provide facilities like load shedding. The controller can be linked to the office computer for items like stock control. These software packs can not only say what is in the stores but also give the exact address of the relevant pallet. Such information is worth protecting. A password security system would be used on sensitive office computer information such as the payroll, and for this there has to be user security on the programable controller. Whilst it may be advantageous for the site engineer to have access to certain levels of the programme to alter parameters such as refrigeration temperature, it could be dangerous if an unqualified person got into the control areas of the programme. A few wrong parts deleted or added could cause chaos within the system costing great sums of money in damaged goods and loss of time.

7.4   Lighting: comfort and safety

Lighting plays two roles in any installation. First, it has to create an environment in which people can work without hindrance and at a high standard of efficiency. Secondly, it has to be available when all other power has failed and there is an emergency such as a fire or a dangerous leakage of refrigerant; in such a situation lighting that people usually take for granted can become a lifesaver.

Whatever type of lighting is used it has to be suitable for operation at very low temperatures; it has to be able to start and run efficiently at $-30°C$, and any lamp failures and blow-outs must be contained, as shattered glass could ruin the stored food that the plant is designed to protect. Plant room areas may require emergency lighting to withstand explosions and fire.

General lighting

The Chartered Institution of Building Services in the UK gives guidance on recommended light levels for a variety of different activities and for different buildings, including cold stores. Large cold stores are generally lit to between 200 and 300 lux; although average illuminance figures are generally quoted, the Code says that care should be taken in lighting the exit and entrance areas to avoid sudden changes of illumination between day and night. Many cold stores operate 24 hours a day, and for these a light source with a warm clear appearance is preferred. This is important with certain foodstuffs such as meat, which have to be visually inspected regularly, and a light source close to that of daylight is essential to ensure that abnormalities can be identified—a coloured light source may be next to useless.

Loading bays can be illuminated to about 150 lux. Higher than this there could be a danger of dazzling drivers as they approach the bay. The edges of the bay must be clearly marked. With the trend towards enclosed loading bays used for sorting and repacking, the lighting levels are often higher. Typically, each loading door will also have stop/go traffic lights so that the driver will not drive off while loading is in progress, and a small spotlight arranged to shine into the back of the vehicle to help with the loading and unloading.

When installing luminaires it is advisable to consider the different types, their light and output performance, life expectancy and reliability. Tungsten is efficient, but does not—in today's terms—last very long in the majority of applications. It is however suitable as a loading bay spotlamp as this is used only irregularly; it is cheap, and quick and easy to replace. On the other hand, in the cold store, the lights are on most of the time and so what is needed here is something that is very cost-effective; the same may apply to emergency lighting which often is on all the time.

Within the cold store environment, the main objective of lighting is cost-effectiveness within the parameters of colour rendering, life expectancy and reliability at low temperatures.

The two most common forms of light source currently used are fluorescent type fittings and high intensity discharge (HID) fittings.

The two most popular types of HID lighting are mercury vapour lamps and high pressure sodium (SON) lamps; others include low pressure sodium (SOX) and metal halide lamps—a derivative of the mercury vapour lamp.

Flourescent tubes are designed to produce various colours and light outputs so the cold store operator can set the colour rendering of the tube according to its replacement cost and energy consumption.

Fluorescent luminaires are generally used in cold store locations where exceptionally good colour rendering is important for repacking or inspection of products or where low ceilings, typically 3 m or less, make the HID lamps unusable because of the glare caused by their very bright light source.

At heights above 3 m, the HID lamps become more economical than fluorescent types, as far fewer luminaires need to be installed for a given level of illumination; running costs are also lower.

Present-day cold stores are mainly constructed with internal heights of 10 m and above and therefore the HID lamp type is the most common. Mercury vapour lamps have been available longer than SON types, but although these have preferable colour rendering (their cold blue/white light is closer to daylight than the light produced by SON lamps), they have a less efficient light output.

In the majority of bulk cold stores, SON fittings are the most popular choice giving a pleasant golden orange/white light and low running costs (Figure 7.3).

The selection of the type of light source to be employed often requires comparison of initial installation and operating costs between mercury and SON lighting. Consideration must be given to positioning within the store. With the introduction of very high racking, including mobile type, the number and location of light fittings is essential in ensuring adequate lighting levels at both the top and bottom pallet, regardless of where a mobile racking aisle has been opened.

Figure 7.3 A typical low profile HP SON T fitting for use in low temperature rooms. When ceiling mounted, it is well clear of the masts of fork lift trucks. The control gear is self-contained and housed in a detachable tray at the rear, connections are made by an internal plug and socket so that the controls of a faulty fitting can be quickly changed, and bench repairs carried out in a warm environment. The fitting is shown, in this photograph, with its clear visor removed for clarity. The visor is essential when the fitting is in operation in the store.

The 24-hour day operation of modern distribution depots with early morning despatches throughout the year makes high demands on the external services within the complex : parking facilities—often with refrigerated vehicle sockets—fuel pump islands, vehicle washes, weighbridges and maintenance workshops. To support this level of activity, higher and higher standards of external lighting are required. No longer is a token level of lighting for security purposes adequate for effective operation of these depots.

Indeed, compared with the minimum recommended illuminance for car parks at 5 lux, many operators now specify external lighting for operational areas of 25 lux, rising to 100 lux in critical areas.

Again, SON type fittings are generally considered the best value for money for external lighting and have the power to deliver the high levels of illuminance now called for (Figure. 7.4).

Emergency lighting

Guidelines for emergency lighting in the U.K. are set out in the British Standard BS 5266: Part 1: 1988. This covers everything from why emergency lighting is required to battery systems, wiring methods, servicing and keeping a log book. The log book is important for fire officers, for example, to check that the system overall meets the stipulated requirements.

The log book must include details of completion certificates, alteration certificates, periodic inspection and test certificates, dates and notes on each

Figure 7.4 A distribution depot during night operation. External illumination is important and, at this site, is provided by 400 W HP SON fittings on the walls of the building and on poles at the perimeter. Twin fluorescent fittings under the canopy provide illumination to the rear of the vehicles. Note the security, closed-circuit, television cameras mounted on the corner of the office building; their operation is enhanced by even, shadow free, illumination.

service, inspection or test, details of defects and remedial action taken and details of alterations to the system. Recalling tungsten lights, these have a short life and as an emergency light in cold stores, they may be expected to be on all the time. Frequently changing the lamps may be time-consuming, but each change has to be logged.

Emergency lights in sub-zero stores cannot be the conventional self-contained type with the battery pack inside because the battery will not operate effectively at such low temperatures—it will not be able to charge and discharge properly. It is usual practice to have a central DC battery system or an AC inverter system supplying remote store fittings. Store fittings are usually tungsten bulkheads, or spot lights for DC systems and fluorescent/tungsten/SON or SOX for inverter systems which are either maintained (on all the time) or non-maintained (come on only in an emergency). The central battery would always be located in ambient temperatures, typically on the loading bay, and give an output of 24–48V DC in case of DC systems and 110–240 V AC in the case of inverter systems.

Some of the primary lighting fittings may be on a separate circuit that will be activated if a generating set is installed to provide back-up power during a mains failure. The problem here is that SON and mercury lights need a restrike period after switching off; this can be anything from 90 seconds to 20 minutes. With an AC inverter, SOX lights would probably be used as these give the maximum light for the minimum input power (Figure 7.5).

7.5   Alarm systems: protecting life and property

One of the important aspects of safety in many plants is that all the devices have to be isolated. This includes equipment such as cooler fans, which are not immediately accessible. In the UK wiring should comply with *IEE Wiring Regulations* (15th edition), including the fitting of earth-leakage circuit-breakers. These are very sensitive to electrical faults and, provided they are correctly fitted, give one of the safest means of protecting personnel against electrocution even with some equipment in very wet areas which may suffer from nuisance tripping. Careful selection of tripping levels can keep this to an absolute minimum.

BS 4434: Part 1:1980 gives general requirements for refrigeration safety. It has quite a broad scope, covering new refrigeration systems, extensions and modifications of existing systems, and used systems which are being reinstalled and operated on a new site. It applies to any refrigeration system in which the refrigerant is evaporated and condensed in a closed circuit.

Fire alarms

The important fire alarm document in the UK is BS 5839: Part 1:1980. The code of practice for installing and servicing fire detection and alarm systems in UK buildings is BS 5839: Part 1:1980. This is a fairly comprehensive

Figure 7.5 View of the mezzanine floor of a low temperature ($-29°$C) distribution store. A high level of lighting for order picking is provided on the mezzanine by the use of fluorescent fittings mounted on suspended trunking. In the background are the top two levels of five-high pallet racking and pallet lifts. The pallet storage area is illuminated by high bay, HP SON fittings over the aisles. Also visible, left of centre and extreme left, are polycarbonate, battery operated, maintained, tungsten emergency lighting over the walkways. Ceiling mounted smoke detectors can also be seen.

document, covering everything from general design considerations to specific details about how to calculate the number of smoke detectors needed.

The first source to be consulted before even starting to build a plant is the local fire officer, who will look at the drawings for the building and give instructions on everything from the electrical system to the construction materials as well as giving advice about which type of alarms and detectors are needed. This can be critical because it may avoid a lot of expense and trouble, trying to modify the building when it is finished. Remember, the local fire officer ultimately has the power to close a plant down if it does not meet the safety requirements.

The basic fire alarm system uses manual break-glass points. These are generally positioned outside the low temperature areas and send signals back to the control panel, which can then shut down parts of the plant. The air movement equipment, such as cooling and ventilation fans, must be shut down because this could otherwise help the fire to spread; also explosive gases have to be prevented from entering the fire zone. The system will also switch on the audio and visual alarms and send signals to , say, the gate house so that the fire brigade can be called.

As well as break-glass systems, the fire officer may insist that heat and smoke detectors are fitted. The plant room is potentially a very hazardous area and the fire alarm installation here has to be intrinsically safe. This means having a remote panel outside the engine room with the detectors in safe boxes.

Some heat and smoke detectors may not work correctly at cold store temperatures. The one principle type that does is called air sampling. This employs an ionization smoke detector located outside the cold store envelope. It takes in a sample of air and measures the ionization of the sample; if smoke or combustion particles are detected in the sample chamber the unit sends a signal back to the main control panel which triggers the alarm system.

There is a problem with audio alarms in a cold store as the background noise of the evaporator fans is often so loud that the alarm may not be heard; hence in some systems the fans will be turned off automatically in an alarm situation.

Also with fork lift trucks it is not unusual for a portable radio to be in the compartment with the driver, and it is therefore essential to have visual alarms as well.

Fire alarm systems may be connected to an extinguishing system. Water sprinklers of course are not suitable for a cold store operating at $-28°C$. Extinguishing systems using halon, an inert gas, may be used for some areas such as the computer room to put out a fire without harming the equipment.

Every fire alarm event should be logged; the fire officer makes frequent random visits and will want to see the log. According to the British Standard mentioned earlier, the log should include not only genuine alarms but also practice, test and false alarms. Faults, service tests and routine attention should be recorded immediately, as should periods of disconnection. If the alarm has been caused by a call point or automatic detector, the location of that device should also be recorded.

Gas and vapour detection systems

Cold stores use various gases as part of the refrigeration system. The gases act as the medium through which heat is exchanged. Typically either ammonia or one of the varieties of freon is used.

Ammonia gas is an extreme irritant that can kill if it is inhaled; it is also explosive at certain concentrations. If there is a leak of ammonia it is vital to have a system to protect personnel and properly. An ammonia detection system can be used to trigger an alarm. This should happen in two stages. The first alarm stage is for when there is only a low level of gas in the atmosphere and the alarm would initiate an audio-visual signal and start the extractor fans automatically. At the second stage, when the gas level is approaching a potentially explosive gas to air mix, the system would shut the plant down completely.

Freons are fluorocarbon based gases and there are a number of different types. They are not directly injurious to people or property, but their density means that they can replace oxygen in the air. Also, under certain conditions, these odourless freons can produce phosgene gas, heat or acid fumes, thus turning a comparatively harmless gas into a killer. Further, freons may cause damage to the ozone layer and restrictions on their use are being introduced. On larger sites, ammonia currently tends to be the more popular refrigerant.

Gas and vapour detection systems will be one of two types. The first checks for potential explosive build-up of gas, and the other for levels that could be high enough to contaminate some of the food products in the store. Food, especially when unwrapped, can absorb small concentrations of ammonia which directly affects its taste and can make it unsellable. It should be possible to detect between 10 and 20 parts per million, but at this level of detection there will be a lot of spurious alarms, such as from hydrogen given off by forklift truck batteries. For this reason most contamination detection systems operate at about 30 to 50 parts per million.

For many years the traditional sensor was the Pellistor, a heated platinum wire in a circuit that gave an electrical response to a change in resistance caused by burning gas. These devices however were not wholly reliable. Semiconductors offer an improvement, as these exhibit massive changes of resistance when exposed to gases other than those found in clean air and their use meets the approval of registration bodies.

The use of microprocessors has heralded an era of remote monitoring for gas leaks. This is already evident in chilled areas in supermarkets and is becoming more common in cold stores. It is possible to have a totally unstaffed refrigeration system with no engineering personnel on site, the gas leak detection system feeding its signal through the microprocessor which in turn would feed it to a remote location.

Whichever system is used it is essential that the gaseous dangers are assessed and the requirements of approving bodies are met before the design of the plant is finalized.

Trapped personnel alarm systems

Almost all cold stores have an easy-to-find call button at ground level by the door that can be operated in an emergency. This has to be run by a battery that is independent of the rest of the electrical system. Early designs of call points incorporated a heater to prevent freezing up, but modern plastics are now available which allow the alarms to operate at very low temperatures. The call point should be well lit, using the emergency lighting, and clearly labelled. When activated, the call button will send an alarm to a central control office and set off flashing lights outside the particular store to make it easier to locate the trapped person. On remote control plants it is possible for the alarm signal to be sent along a telephone line to alert the central controller.

Security systems

Cold stores may store goods with an extremely high value and—like any large building with few people about—present a potential security problem, although they do have the advantage that they operate 24 hours a day in many cases.

The easiest entrance for an intruder would probably be the emergency exits at the back of the store—these are generally unstaffed and out of sight from the rest of the site. Closed-circuit television is the most common method of surveillance for areas such as these, and can operate at a low level of illumination, say 1 to 2 lux, or even when equipped with infra-red sensors in total darkness. The problem is that a security guard staring at a bank of screens all day can easily miss something, especially as months or years can go by with nothing happening. It is thus wise to link the television circuit to a video recorder so that any break-in would then be captured on film.

The emergency exits, and indeed all doors, should have standard alarm systems fitted. These should either be on their own battery system or linked to stand-by power: it is not uncommon for a power failure to bring the intruders out.

7.6   Cold store installations: specific requirements

Refrigerated vehicle sockets

When frozen goods are transported, large refrigerated vehicles are used which are powered from the truck itself. When the truck arrives at a site it may not be possible to unload immediately and the trailer may be left standing in the yard, perhaps for a whole weekend. For this reason, cold stores tend to have a large number of special sockets in the yard that can be used to power a vehicle's refrigeration system.

For small trucks, 16 A three-phase sockets are normally adequate, but for larger trucks sockets can be up to 32 A three-phase. They are normally five-pin, triple-pole neutral and earth (TPNE) or just triple-pole and earth (TPE) sockets. Some of them have phase rotation switches to compensate for different wiring systems on different trucks. Almost all are earth leakage protected sockets and are mainly interlocked with the plug heads.

The requirement for large numbers of such sockets is often not anticipated when initially designing a cold store's electrical installation and, since each socket can draw in excess of 20 A three-phase, the implications for the electrical capacity of the cold store are obvious. To avoid the cost of building additional electrical capacity into the store, some cold store operators let the trailer vehicles use their own diesel power motors but this can lead to problems of noise pollution, especially if the plant is near residential areas, and in any case the economics of diesel engines versus electric motors favours the use of electrical sockets (Figure 7.6).

Figure 7.6 A typical vehicle parking area with power sockets for the truck-mounted refrigeration units. Vehicles reverse onto 'bump kerbs' to either side of the island and plug into socket outlets. Each socket is enclosed in a weatherproof housing, mounted on a robust post, and protected by a substantial hood. Each socket outlet has electric shock protection provided by an earth leakage circuit breaker. Feeds to the sockets are provided from a local distribution board mounted in the weatherproof housing in the foreground. This housing has an anti-condensation heater and integral lighting.

Frost heave protection

Cold stores can affect the environment and in particular the ground underneath them. A large block at $-25°C$ or below is difficult to isolate. If no care is taken an 'ice lens' may build up below-ground and damage the foundations of the building or cause 'frost heave'.

Large plants generally have heated wastepipes below the floor insulation to prevent the ground freezing. For smaller plants, this is often not practical and electrically heated mats are used as an easier alternative.

The under-floor mats can either be mains-powered or run at low voltage via a transformer. The obvious advantage of using mains power is that the transformer is not needed. The disadvantage of using mains power is that should a wire break, the mains could be transferred to the steel-work of the building. This is relevant because there is a long-term reliability problem with mains heaters and it is not uncommon for there to be a separate heater in the core of the cable to act as a back-up.

The low-voltage system provides an electrically energized source of heat as wires laid on top of the site concrete. Over this grid, or heater mat as it is called, is floated a sand and cement screed. On top of this will be the vapour sealing and floor insulation. The core of the system is a high-grade stainless steel wire working at temperatures only a few degrees above the temperature of the concrete slab to which it is fixed. The wires are energized from the low voltage output of a step-down transformer. Such a low voltage system can be operated without insulation, but sheathing on the wire does give added protection to the system and makes it easier to terminate individual heating elements.

Large, low-temperature chambers often have exposed steel columns which can transfer the sub-zero temperature down through the cold store floor to the ground; these too may need to be heated. A similar problem may occur where columns pass through an insulated ceiling into a roof void; sometimes heaters are used here to avoid condensation and frost build-up on the columns in the void.

Door opening/closing systems

For every second that a cold store door is open, the store warms slightly and it costs money to cool it back down. Despite this, the most common method of opening and shutting doors is with a pull switch. This has obvious problems; for example a fork lift truck driver picking up a pallet near the door may be tempted to leave the door open during the visit. One solution may be by putting a timer on the door so that after a few seconds the door shuts automatically. This type of system is normally backed up either by a plastic strip curtain behind the door to stop some of the air change, or by blowing air across the opening to cut down air movement into the store (a lot of the larger High Street shops use this method). Neither is a substitute for any method that would close the door more quickly. It is not practical to set the timer at a lower setting because different drivers with different types of load take different times to negotiate the exit.

One very useful device is to install induction loops in the floor that create a magnetic field which changes when a metal object, like a fork lift truck, crosses it; this creates a signal that can be used to open or close the door. Some operators even use a complex double-loop system that will discriminate between a driver going past the door and one actually trying to go through the door. It is also possible to use a radio transmitter and receiver system which lets the driver operate the door from the cab (see also Chapter 10).

Stand-by power

The major application for stand-by power is not the low temperature refrigeration system itself: this normally has enough 'flywheel effect' within the system that it can survive for some hours with no power without warming to an unacceptable level. Sometimes low temperature stores can survive for days if lights are off and doors closed. The exception would be chill stores, where the temperatures are higher and the products in store are temperature-sensitive. Temperatures in chill stores can change quickly and some form of stand-by power would probably be needed very quickly to continue operation after a power failure.

The best known type of stand-by power is the generating set. This would probably be linked to the lighting and other electrical safety alarms and devices, and to the chill store if one existed.

A generator set which has been specifically designed, produced and installed to meet the precise requirements is often a much better solution than an off-the-shelf model that is 'about right'. A generating set is at its most efficient when operating close to the top of its range, so tight matching is important. Further points to note are that generating sets are noisy and it may be necessary to put soundproofing materials around it, and that different sets have different start-up times.

A generating set normally takes about 10 seconds to be run up to full speed. It is then introduced to the loads in a staggered fashion, either manually or automatically. If it is done automatically, predetermined priorities of loads, such as computers, lighting of offices, must be established. Any delay even as short as 10 seconds would be disastrous for a computer system—even a fraction of a second of loss of power can kill and lose all the programs being run at the time. In this case the answer is to have some form of uninterruptible power supply.

An uninterruptible power supply (UPS) comprises four elements—the rectifier and battery charger, storage battery, inverter and by-pass and transfer switch—which are set up between the mains and the load. All connected systems are powered by the battery and the mains is used to keep the battery charged. This has the added advantage that it filters out spikes and other interference and can keep the power running long enough for the generating set to come on line. During normal operation, the rectifier converts the AC input power to DC which supplies the inverter and keeps the battery fully charged. The inverter then converts the DC to a regulated and clean AC to send out to the computer. If there is a power cut, the inverter draws its power from the storage battery. When the power comes back on the rectifier automatically goes back to its recharging role. Single units are available from under 1 kVA and up to 400 kVA. Parallel, redundant and multimodule configurations can take the capacity up to 2.5 MVA or higher.

Energy saving

More than 40% of the running costs of a cold store can be spent on energy and therefore any means to save energy are welcomed by cold store operators.

The most obvious method, as mentioned earlier, is to run the refrigeration plant mostly at night during off-peak electricity times. Further, a generating set could be used to run the plant during peak tariff periods. If running on a maximum demand tariff, a computer-controlled system should be used to switch off nonessential circuits as usage approaches the maximum demand figure. The system could do this either automatically or by sounding an alarm to warn the operator that systems will need to be shut down. The operator can then manually shut down certain systems.

The designer of the refrigeration system has to take into account the heat generated by the lighting system: the lights put heat in, and it costs money to get that heat out again. Automatic door closing such that doors are open for a minimum time when fork lift trucks go through is also desirable. The longer doors are open, the more heat gets into the cold store.

The efficiency of power generation equipment and the main compressor motors on the plant is improved if the power factor of the generator is as near as possible to unity. This can be achieved by using capacitors installed near the load. There are a number of systems on the market, ranging from fixed value banks sized to particular loads to multistep units which modulate in or out of circuit to retain an acceptable power factor over various production load patterns.

7.7 Conclusion

There is no doubt that the biggest revolution in cold store control, that of the change from electromechanical relays to the modern computer and programmable controller systems has already happened in most present-day installations. As well as giving cost savings and ease of use advantages, these systems are flexible and can be expanded easily with few problems.

This flexibility is becoming more and more important as requirements of cold stores come to change rapidly. At one stage the plant may be used just for bulk storage, but later may become a busy distribution centre with goods coming and going all the time—more vehicle sockets may be needed, doors may open more often putting greater demands on the refrigeration, and capacity may have to be increased.

All this is combined with a trend towards greater economy through energy management and power factor correction. Safety regulations are now more stringent and the risk of accident in a busy distribution depot is higher than in a plain storage plant.

The electromechanical relays of earlier installations cannot handle change and they have been replaced by the silicon chip with all its advantages. In cold stores as in the rest of industry, microelectronics has shown that it has a role to play.

Acknowledgements

The author would like to thank the Electricity Council, London Electricity Board, Yorkshire Electricity Board, British Standards Institution, Star Refrigeration Ltd, Electronic Devices Ltd, APV Hall International Ltd, Lowheat Ltd, Natalarm Systems Ltd, Guardian Fire Detector Systems Ltd, MPL Power Systems PLC and Petbow Ltd for information they have provided which has helped in producing this chapter.

8 Racking systems

R. SHOWELL

8.1 Introduction

A racking system is one constituent part of a warehouse system and can be fully defined only when all the other parameters of a warehousing operation are considered. It is dependent on what the user requires from his operational unit.

At one end of the scale is the system that is 'all things to all men', and at the other end is a system dedicated to one customer with a fixed long-term requirement. Neither of these extremes is generally acceptable, and therefore the warehouse system becomes tailor-made to suit individual customer requirements.

The more flexible the system is to cope with changing demands, the more effective it is. There is, however, always a cost penalty for this flexibility, be it in higher initial capital cost or lower operating revenue.

This chapter describes the racking systems currently available, and compares a range of layouts.

8.2 Standards

Racking systems are covered by different standards and codes of practice in different countries of the world. In the UK, the industry generally conforms to the Storage Equipment Manufacturing Association (SEMA) codes of practice for the design and installation of racking systems. This does not, at present, cover all the systems described. However, manufacturers of this equipment should also follow the relevant standards. Probably the three most important UK standards are BS 449, which covers racking system materials, BS 2629 for pallets, and BS 5750, which embraces excellence of quality, design, and manufacture.

8.3 Basic unit of storage

Before any racking system can be considered, the basic unit of storage must be defined. This is shown diagrammatically in Figure 8.1.

Most storage operations have to take into account a number of options

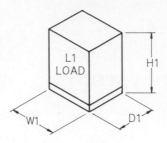

Figure 8.1   The unit of storage. H1 = height including pallet; D1 = depth; W1 = width on entry face; L1 = Load.

before arriving at a workable unit of storage. For a racking system to be workable, it is necessary to analyse the 'unit of storage' far more precisely. The following further information must be considered:

- Load to be stored
    - Is it boxed/bagged/loose/crushable/dense?
    - Does it overhang the pallet? If so, by how much in each direction?
    - Will the load settle in storage and lean outside the basic pallet dimensions? If so by how much?
    - Does the load require restraining in some way (stretch-wrapping or banding)?
    - How heavy is the load in its most dense form?
    - Is the load to be 'broken down' in storage area or off the premises?
    - What is the period of time that the load is to be in store?
- Storage platform (pallet)
    - Is the pallet, for use within the plant, used within the UK, within the EEC, or worldwide?
    - Are the pallets reusable or one-way?
    - Do the pallets have to comply with standards. e.g. UK/AFFP/NCSF and BS 2629:1967?
- Storage area
    - What are the internal dimensions of the room to be used?
    - What are the positions and use of the doors? Are they for mechanical handling equipment, pedestrians, or other users?
    - What are the future developments which might restrict the layout of the racking systems?
    - What is the operating temperature?

As the racking and mechanical handling system is dependent on the load to be stored and the storage platform, it can influence the storage area. It is, therefore, important that the above questions and statements are answered before the racking system or systems are to be considered.

The load is the parameter which dictates everything. This must be defined,

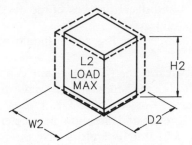

Figure 8.2 Revised unit of storage. H2 = maximum height, including pallet; D2 = maximum depth; W2 = maximum width on entry face; L2 = maximum load.

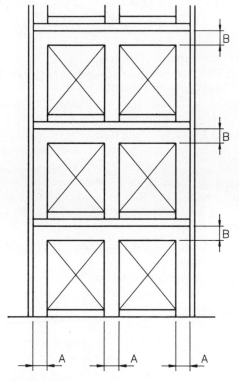

BEAM HEIGHT FROM FLOOR (mm)	A (mm)	B (mm)
0 – 3000	75	75
3000 – 6000	75	100
6000 – 9000	75	125
9000 – 12000	75	150

Figure 8.3 Front elevation view of pallet location in standard racking system. A = clearance between adjacent pallets or loads, whichever is the greatest, and the pallet/load or rack structure; B = clearance between underside of pallet support beam and top of load at its maximum when handling equipment is *not* fitted with automatic height selection equipment.

and all the advantages and disadvantages of the deviations from the norm should be fully understood.

The unit of storage shown in Figure 8.1 should now be redrawn as in Figure 8.2.

From this revised load size, the racking system can now be determined; it should be determined using the codes of practice prevailing in the country in which the system is being installed. In the UK, the code is defined by SEMA. The code lays down maximum and minimum parameters where applicable. At least three major UK cold store operators have their own standards which allow for an improvement on either the maximum or the minimum.

The standard is summarized in Figures 8.3 and 8.4 which show the typical dimensions for a static pallet racking system where the handling equipment is not fitted with automatic horizontal or vertical positioning equipment.

The following points should be noted when considering Figures 8.3 and 8.4:

- Reducing the clearance between the pallet/loads will increase the incidence of damage to both product and racking. It will also increase the pallet placing and retrieval time
- Mixing pallet types within a racking system will result in a higher incidence of pallet failure
- Using sub-standard pallets will result in pallet collapse, leading to product and racking damage
- The type of storage facility that is required must be planned together with all the financial implications

8.4 Storage concepts

The following are the racking sytems in use in temperature-controlled storage today.

- Block stacking
  —free standing
  —with converter sets
  —post pallets
- Pallet racking
  —static adjustable pallet racking (APR)
  —narrow aisle adjustable pallet racking (NAPR)
  —drive-in/drive-through pallet racking
- Live storage
  —first in, first out (FIFO)
  —last in, first out (LIFO)
- Double deep static racking
- Powered mobile
  —adjustable pallet racking (PM APR)
  —narrow aisle pallet racking (PM NAPR)

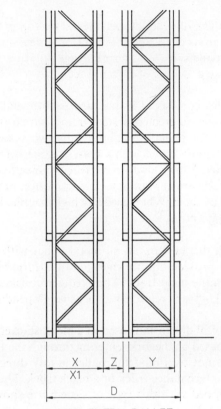

2 WAY ENTRY PALLET

X	Y	Z	D (X1 + Z + X1)
750	600	100	1600
900	700	100	1900
1000	750	150	2150
1200	900	150	2550

4 WAY ENTRY PALLET

X	Y	Z	D (X1 + Z + X1)
750	700	75	1575
900	800	75	1875
1000	900	100	2100
1200	1100	100	2550

Figure 8.4 End elevation view of pallet location in standard racking system. X = overall depth of pallet; X1 = overall depth of load; Y = dimension over outside of pallet beams; Z = clearance between back to back of pallets or loads; D = dimension over front to front of pallets or loads. Quoted dimensions of X, Y, Z and D are in mm.

Block stacking

Free standing   Pallets of product are stacked on top of each other in rows, usually no more than six deep. Suitable only for large numbers of identical-use loads on first class pallets in a long-term storage situation. Probably limited to three or four high, e.g. drums of frozen juice.

Converter sets   This consists of a standard pallet usually 1 m × 1.2 m four-way entry onto which is erected a steel frame, which provides rigidity for supporting subsequent pallets. A labour-intensive system, as it is usually necessary to fit an empty pallet with a converter set before the load is hand balled onto the pallet. Allows loads to be stored four or five high. It is essential that the converter sets are kept in first class condition as the whole weight of the stack is carried by them. When used five high in a block stacking situation, it produces very high point loads on the floor.

Post pallets   With this system, an all-metal pallet with corner legs is used (Figure 8.5). These provide a more rigid load platform for storing five high. Used extensively in the meat trade. They require considerable storage space when not in use. They are relatively expensive, depending on strength and finish.

A plan of the typical dimensions used in a block stacked system is shown in Figure 8.6. This is based on the forks entering through the 1.2 m face. It assumes that the load does not overhand the pallet in any direction. Loads can be handled with a counterbalance truck or a reach truck. The typical dimensions of a chamber equipped with this type of system are shown in Figure 8.7.

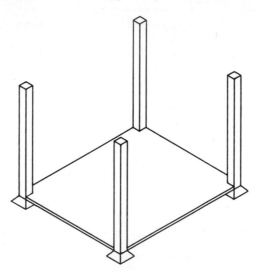

Figure 8.5   Diagram of a post pallet.

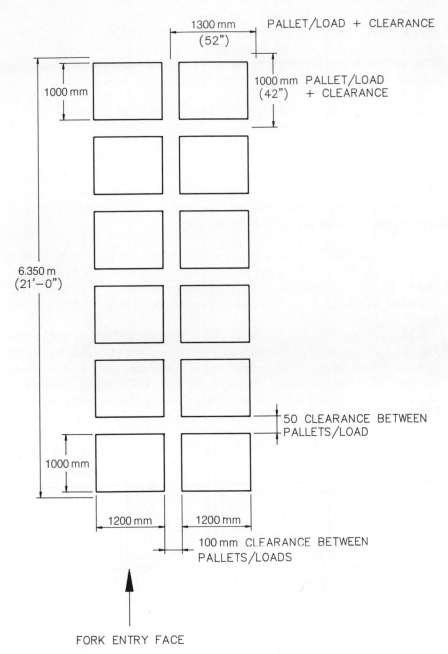

Figure 8.6 Typical dimensions of a block stacked system (plan view).

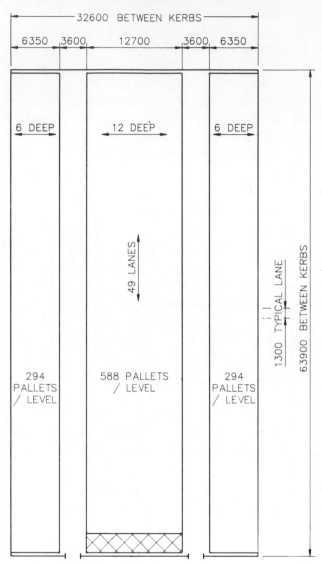

Figure 8.7 Typical block stacked installation for use with a counterbalance truck handling the pallet on the 1200 mm face. Hatched region indicates area to be left clear so that truck can move from one aisle to another in the event of door failure. Dimensions are in mm.

Pallet racking

Static adjustable racking (APR) Pallets are supported, usually in twos or threes, side by side on horizontal beams. As each load is free standing, there is only a requirement to ensure that the load is stable on its own pallet. This system provides instant access to any pallet, is cheap to install, but very expensive in the amount of space required for each pallet. Can be adjusted vertically to accommodate different pallet heights.

The system usually consists of two components (Figure 8.8):

- *End frames*, which can have either welded or bolted bracing. Both types are available and should comply with SEMA requirements. The upright struts have pre-formed holes or slots at regular intervals up their outer faces. These holes or slots are usually pitched at 50/75/100 mm and provide the fixing and method of adjustment for the pallet support beams.

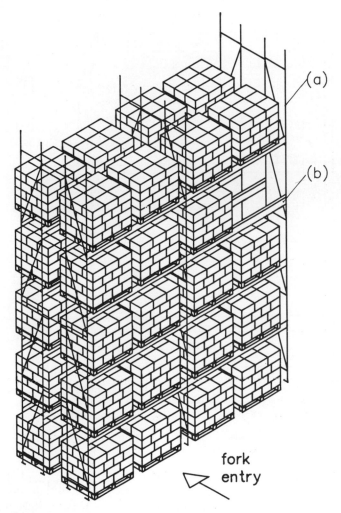

Figure 8.8   Static adjustable pallet racking. (a) End frame; (b) pallet support beam.

- *The pallet support beams.*   These are in pairs and are usually pre-formed and fitted with end connectors which locate in the holes/slots.

A room using this system would need an additional 1 m clear height when compared to an equivalent five-high block stacked system. Pallets are

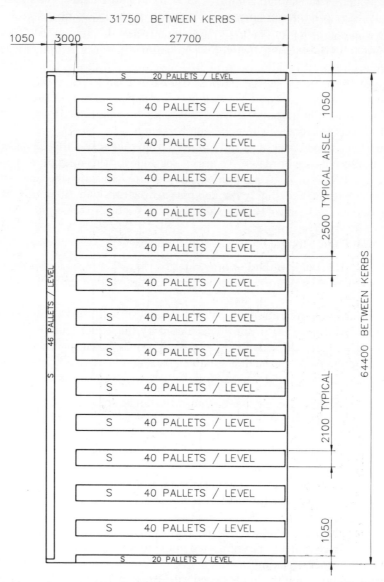

Figure 8.9 Typical static pallet rack installation for use with a reach truck handling the pallet on a 1200 mm face (plan view). Capacity with five levels is 3030 pallets. Dimensions are in mm.

normally handled with reach or counterbalance trucks. The first or ground floor pallet can be handled with a pallet truck.

A plan view of a typical static APR system is shown in Figure 8.9. Fork entry is through the 1.2 m face and the load does not overhang the pallet. The aisle width assumes the use of a reach truck as this normally requires less space than a counterbalance truck for manoeuvring.

Narrow aisle adjustable pallet racking (NAPR)   Pallets are supported in the same manner as in APR. The load requirements are also the same, as is the basic design of the racking structure. The major difference is that the aisle between each run of racking is up to 30% narrower than for APR. The cost penalty for this is that a special-purpose truck is needed to place and retrieve the pallets from the racking. The truck moves much faster within the racking than a reach or counterbalance truck but is not normally used outside the racked area as it becomes slow and cumbersome.

Pallets have to be brought to and retrieved from the racked area by a feeder truck (counterbalance or reach). An area is set aside for this operation. This is known as a P&D (place and despatch) station and is usually the first pallet stack in each run of racking. It can be approached from two directions: from the end of the rack by the feeder truck and from the side or storage face by the narrow aisle truck.

A typical pallet storage sequence is as follows:

- A pallet is brought into the storage area by the feeder truck and placed on any vacant P&D station corresponding to the aisle that the pallet is to be stored in.
- The narrow aisle truck then picks the pallet up with forks entering the pallet face that is 90° from the face used by the feeder truck.
- The truck with the incoming pallet then moves down the narrow aisle placing the pallet in the first suitable position on either side of the aisle.
- The reverse procedure is used for pallet despatch out of the storage area.

NAPR is more expensive to install than APR but has better cube utilization. It requires specialized handling equipment which will dictate that long narrow chambers are the most cost-effective. On average a store has to be 150 mm higher than that required for an equivalent APR system and the floor has to be laid to ± 2 mm in 2 m.

A typical plan of an NAPR system is shown in Figure 8.10. Fork entry is through the 1 m face and the load does not overhang the pallet.

Drive-in drive-through pallet racking   This system uses a form of racking which allows each pallet to be supported individually but gives a similar cube utilization to block stacking.

Drive-in dictates that the loads stored will be last in first out (LIFO). Drive-through enables a first in first out (FIFO) operation to take place.

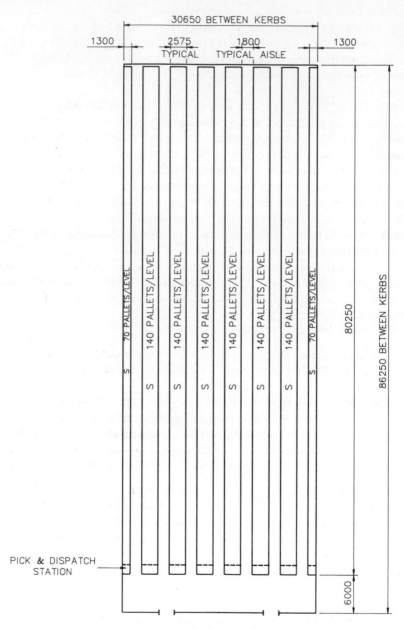

Figure 8.10 Typical narrow aisle static rack installation for use with a narrow aisle truck handling the pallet on the 1000 mm face (plan view). Capacity with five levels is 4900 pallets, including 70 pick and dispatch. Dimensions are in mm.

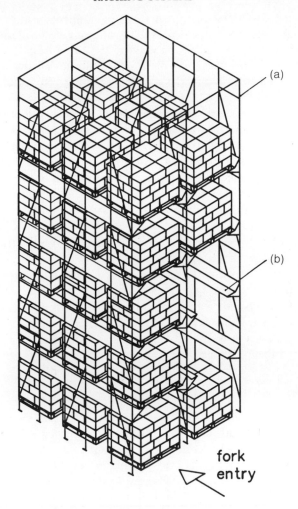

(a)

(b)

fork
entry

Figure 8.11 Drive-in–drive-through pallet racking. (a) End frame; (b) pallet support rail.

The system uses racking uprights fitted with pallet support rails cantilevered from the side of the uprights (Figure 8.11). These rails run from front to rear and form the required storage levels. The uprights are fixed to the floor and tied above the top pallet level. The structure is also braced above the top pallet level. Lanes are therefore formed down which a truck can be driven to access each level of pallets. All levels in each lane have to be filled or emptied at the same time, which, in turn, means that each lane must contain the same product.

The difference between drive-in and drive-through is that with drive-in, each block can only be accessed from one side, whereas drive-through can be accessed from both sides with no structural impediments in any of the lanes. As drive-through requires a higher level of strength to be derived from the steelwork above the top pallet, it is therefore less cost-effective than drive-in. A simple solution is to place two drive-in installations back to back when access is required from both sides and FIFO is not important.

Either a counterbalance or a reach truck can be used with this type of installation, but care is needed to ensure that the truck will pass between the pallet support rails and the frame fixing feet. In some instances modifications can be made to the trucks to enable them to operate satisfactorily.

As the load on each pallet is transmitted through the outermost pallet members on two parallel sides only, it is imperative that the pallets are well made and in very good condition.

The floor level within the drive-in installation must be to within $\pm$ 3 mm in 1.5 m.

As the load has to be lifted to the correct level before the truck enters the racks, this will slow the throughput rates.

Drive-in is more expensive than APR to install, but has a cube utilization of about 80% of block stacking.

Some drive-in systems will fit into the same height chamber as a NAPR system (i.e. 1–1.5 m higher than a blocked stacked system).

A typical plan of a drive-in installation is shown in Figure 8.12 with the fork entry through the 1.2 m face.

Live storage

First in, first out (FIFO)   Live storage systems have the same lane configuration as drive-in racking, except that each level of pallets is supported on a shallow sloping roller, wheel, or conveying system. Pallets are placed on at one end, roll down the sloping track to accumulate at the other.

The type of rollers or wheels will largely depend on the weight of the loads to be stored and the condition of the pallets in use.

Some systems will require special slave pallets which are captive within the system. The slave pallets have to be recycled from the output end to the input end by handling equipment.

On longer lane installations, pallet separating devices should be filled to reduce the line pressure. Retardation devices will also be needed near the output end to control the forward speed of the first two or three pallets placed in each lane.

The floor level need only be to within $\pm$ 3 mm in 1.5 m. However, the structure should be fitted with adjusting feet under each end frame upright to ensure that the designed slope is maintained.

The cost per pallet will be of the order of two to three times as expensive as an equivalent drive-in system.

As the pallet lanes have a 3–6° slope between input and output end, a higher

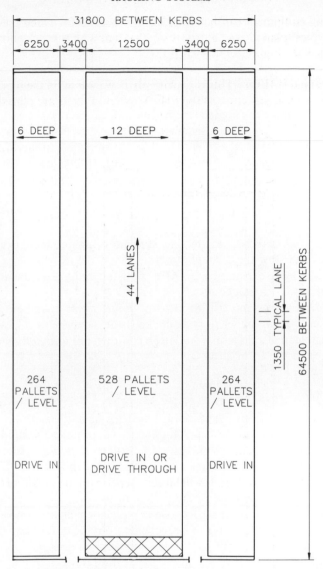

Figure 8.12 Typical drive-in racking installation for use with a reach or counterbalance truck handling the pallet on the 1200 mm face (plan view). Capacity with five levels is 5280 pallets. The centre block can be either drive-in or drive-through; the outer blocks are always drive-in. Hatched region indicates area to be left clear so that truck can move from one aisle to another in the event of door failure. Dimensions are in mm.

chamber will be needed than any of the previous systems described. This increase in height is a function of the depth of the system. For a 12-pallet-deep installation, the height of the chamber for a five-high system will need to be 2.5–3.0 m higher than an equivalent block stacked system.

Handling equipment can be either counterbalance or reach trucks.

For a typical plan view of a live storage system with the fork entry through the 1 m face see Figure 8.13.

Last in, first out (LIFO) This is a relatively new system on the market, which has the same basic structure as the FIFO system. Pallets are placed either on

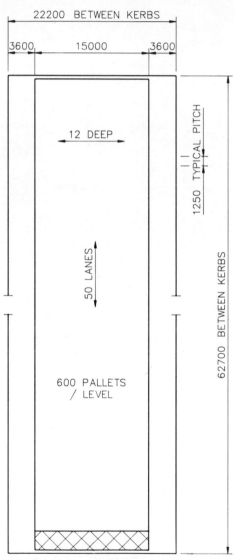

Figure 8.13 Typical FIFO live storage system for use with a counterbalance truck handling the pallet on the 1000 mm face (plan view). Capacity with five levels is 3000 pallets. Hatched region indicates area to be left clear so that truck can move from one aisle to another in the event of door failure. Dimensions are in mm.

rollers, guides, or platens and pushed away from the input face by the next pallet.

Because the pushing force is provided by the handling equipment, there is a limitation on the depth of the system. It is likely to be up to four deep for 1000 kg pallets and up to eight deep for 300–500 kg pallets.

Cost indications, building heights and handling equipment are the same as

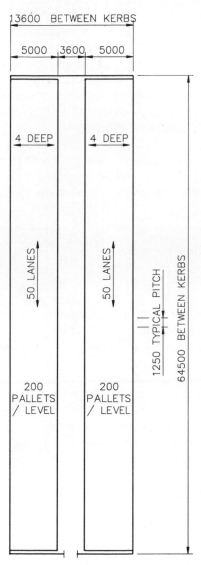

Figure 8.14 Typical LIFO live storage system for use with a counterbalance truck handling the pallet on the 1000 mm face (plan view). Capacity with five levels is 2000 pallets. Dimensions are in mm.

the FIFO system. However, there is a floor space saving as a common access gangway serves for both input and output operations.

A typical plan of a LIFO live storage system is shown in Figure 8.14.

Double deep static racking

This is basically the same racking structure as APR except that the pallets are stored two deep from any access aisle, i.e. 50% of all the pallets can be accessed immediately. The main advantage over conventional static racking (APR) is that 30% more pallets can be stored in a given area.

There are two main disadvantages. First, a special double reach attachment is needed on the FLTs used. Few trucks can be modified in this way without considerable loss of load carrying capacity. Secondly, for truck stability reasons, the truck outriggers must pass under the bottom pallet. This means that all pallet levels are supported on beams.

The overall height of a store has to be increased by 500 mm when compared to an equivalent static installation. The ground floor (first pallet) cannot be handled by a pallet truck.

A typical double reach installation handling on the 1.2 m face is shown in Figure 8.15.

Powered mobile racking

Powered mobile adjustable pallet racking (PM APR) The racking design concept is the same as APR. The major difference is that all but one of the access aisles between the racks are eliminated. The racks are mounted on power-operated bases which move them to the left or right giving immediate access to every pallet. Rails are set in the floor which has to be laid to ± 2 mm in 2 m.

Safety bars/skirts are provided down each side of the mobile racks to stop them should they meet an obstruction.

Control of the racks ranges from simple push-button stations on the ends of the racks, to control systems mounted within the heated cabs of FLTs. Considerable development work has been done over in recent years to see that the latter systems meet all the necessary safety requirements, ensuring that racks are not moved until the open moveable aisle is inspected for obstructions.

This system gives immediate access to any pallet. The increase in storage capacity over a static pallet rack system is likely to be better than 85% for a given storage area. An other advantage is that the FLT travelling distance is reduced by up to 50%.

If a sequential pallet placing/retrieving documentation system is used, then the time taken to move the racks is insignificant.

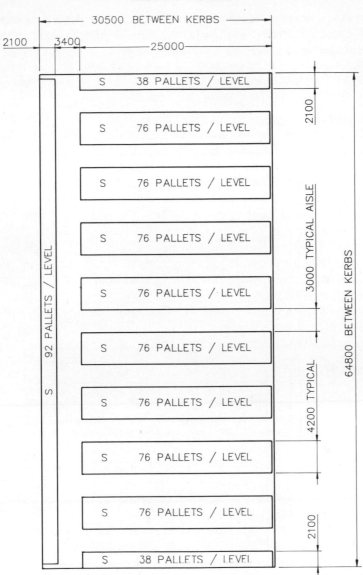

Figure 8.15 Typical double deep static rack installation for use with a specially adapted reach truck handling the pallet on the 1200 mm face (plan view). Capacity with five levels is 3880 pallets. Dimensions are in mm.

A reach truck is usually employed with this type of installation. It is also common practice to make the moving access aisle wider than normal (3.0–3.4 m). It has been proved that by adding 300–600 mm to the minimum aisle width recommended by reach truck manufacturers, the incidence of product and racking damage is reduced dramatically with a corresponding increase in the work rate from each truck. The luxury of a wide aisle can easily be justified with this system in a controlled-temperature store.

Pallet movements of in excess of 30 pallets per hour per reach truck are

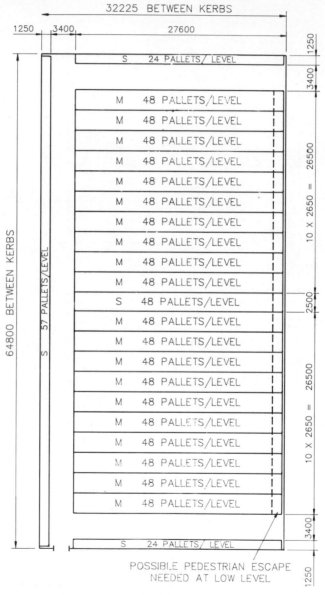

Figure 8.16 Typical powered mobile rack installation serviced by two reach trucks handling the pallet on the 1000 mm (plan view). Capacity with five levels is 5523 pallets.

achieved in some modern cold stores. Over 65% of new cold stores are now built incorporating powered mobile racking as the storage system.

The layout of a store with this system is critical in order to achieve the optimum cube utilization along with the throughput required. *The building should be fitted around the storage system and not the other way round.*

Order-picking can be carried out from the first level, but the layout has to be very carefully considered. Mobile systems are ideal for medium to slow-moving lines or as a random access back-up storage system for a fast moving order-picking area.

This system requires a similar height chamber to that for a drive-in installation.

It is more common to handle the pallet on the 1 m entry face than the 1.2 m face, as this produces further equipment economies coupled with a greater number of pallet positions available in any open aisle. A typical layout for a power mobile racked system using reach trucks handling pallets on the 1 m

Figure 8.17 Cold store equipped with power mobile racked system.

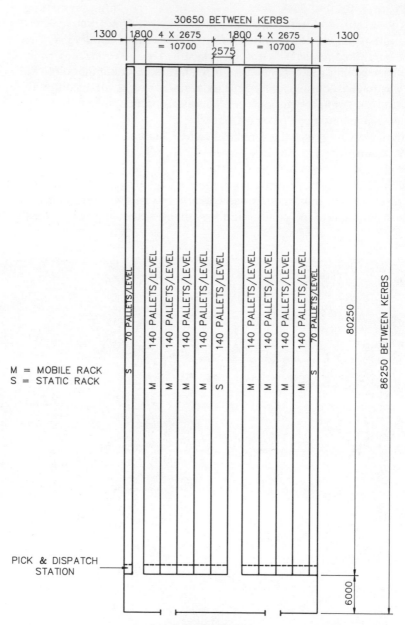

Figure 8.18 Typical narrow aisle mobile rack installation serviced by two narrow aisle trucks handling the pallet on the 1000 mm face (plan view). Capacity with five levels is 7000 pallets, including 70 pick and dispatch positions.

face is shown in Figure 8.16 and a typical cold store so equipped is shown in Figure 8.17.

Powered mobile narrow aisle adjustable pallet racking (*PM NAPR*)   In basic concept, this system is the same as PM APR. The major differences are, first, that the moving aisle is only 1.6–1.8 m wide, to accommodate the use of a narrow aisle truck; secondly, guiding surfaces are incorporated within the mobile base design for the guidance of the trucks; and thirdly, P&D positions are placed at one end of every mobile rack.

All the controls for initiating the movement of the racks are mounted on the FLT. Again safety features are incorporated to ensure that the racks cannot be moved until the open aisle to be moved is proved clear.

The system is best installed in long narrow chambers with 60 pallets or more per level per rack face. Pallet movements in excess of 45 pallets per hour per narrow aisle truck are achievable with this system.

Figure 8.19   Cold store equipped with powered mobile narrow aisle racking.

Table 8.1 Summary of the physical of capacities the layouts (theoretical).

	Block racking	Static racking	Narrow aisle	Drive-in, drive-through	Live Storage FIFO	Live Storage LIFO	Double deep static	Powered mobile	Powered mobile narrow aisle
No. of pallets stored	5880	3030	4900	5280	3000	2000	3880	5523	7000
Cubic volume of store (m³)	17700	19424	25510	19895	15310	9380	19266	20270	25245
Volume per pallet (m³)	3.01	6.41	5.21	3.77	5.10	4.69	4.96	3.67	3.61
Rating based on volume	1	9	8	4	7	5	6	3	2
Floor area per pallet including gangways (m²)	1.77	3.37	2.70	1.94	2.32	2.13	2.55	1.89	1.86
Rating based on area	1	9	8	4	6	5	7	3	2
Handling equipment costs compared to counterbalance trucks used in block stacking		1.5	3.5	1.0 to 1.5	1.0	1.0	2.0	1.5	3.5

Note 1   Includes all gangways, aisles, racking structures and recommended pallet clearances
Note 2   Excludes any additional height needed for air circulation above top pallet
Note 3   All layouts are five high
Note 4   All layouts are based on 1000 mm (40″) × 1200 mm (48″) × 1675 mm (66″) load size

Table 8.2 Summary of the physical capacities of the layouts (practical).

	Block racking	Static racking	Narrow aisle	Drive-in, drive-through	Live Storage FIFO	Live Storage LIFO	Double deep static	Powered mobile	Powered mobile narrow aisle
Theoretical no. of pallets stored	5880	3030	4900	5280	3000	2000	3880	5523	7000
Loss due to type of storage system (%)	44	10	10	41	39	39	25	10	10
Practical no. of pallets stored	3292	2727	4410	3115	1830	1220	2910	4970	6300
Volume per pallet (m³)	5.38	7.12	5.78	6.39	8.37	7.69	6.62	4.08	4.01
Practical rating based on volume	3	7	4	5	9	8	6	2	1
Floor area per pallet including gangway (m²)	3.16	3.75	3.00	3.29	3.80	3.49	3.40	2.10	2.07
Practical rating based on area	4	8	3	5	9	7	6	2	1

Note 1 Includes all gangways, aisles, racking structures and recommended pallet clearances
Note 2 Excludes any additional height needed for air circulation above top pallet
Note 3 All layouts are five high
Note 4 All layouts are based on 1000 mm (40″) × 1200 mm (48″) × 1675 mm (66″) load size

Table 8.3 The relative merits of different racking systems. Asterisks denote rating out of 5.

	Block Stacking			Static racking APR	Narrow aisle NAPR	Drive-in	Drive-through	Live storage		Double deep static	Powered mobile PMAPR	Powered mobile narrow aisle PMNAPR
	Free standing	Converter sets	Post pallets					FIFO	LIFO			
Cube utilization												
Theory	*****	*****	*****	*	***	****	****	***	***	**	****	*****
Practice	***	***	***	*	***	***	***	**	**	**	****	*****
Pallet accessibility	*	*	*	*****	*****	**	**	**	***	***	****	*****
Stock rotation	*	*	*	*****	*****	*	****	*****	**	***	*****	*****
No. of product lines	*	*	*	*****	*****	**	**	***	***	***	*****	*****
Lack of product damage	*	***	***	**	****	**	**	****	****	**	***	****
Inventory control	*	*	*	*****	*****	***	*****	*****	*****	***	*****	*****
System type	LIFO	LIFO	LIFO	RA	RA	LIFO	FIFO	FIFO	LIFO	LIFO	RA	RA

Note   RA = Random Access

Table 8.4 Economic factors for each racking system.

	Block Stacking			Static racking APR	Narrow aisle NAPR	Drive-in	Drive-through	Live storage		Double deep static	Powered mobile PMAPR	Powered mobile narrow aisle PMNAPR
	Free standing	Converter sets	Post pallets					FIFO	LIFO			
Area of store covered (%)	80	80	80	39	60	75	75	70	70	55	85	88
Increase in store height over 5 high block stacking	0	0	0	1000 mm 36″	1150 mm 42″	1200 mm 48″	1200 mm 48″	2500 mm 100″	2500 mm 100″	1250 mm 52″	1200 mm 48″	1200 mm 48″
Suggested minimum depth of pallet runs	6	6	6	10	36	4	8	6	—	10	20	30
Suggested maximum depth of pallet runs	6	6	6	30	80	10	20	20	4	30	60	80
*Cost range/pallet stored (£) Low	good pallets and load	37	43	19	25	32	36	85	85	24	54	65
High		68	95	23	30	43	48	140	140	31	89	110
Suggested floor-to-top of top pallet (1675 mm, 5′6″ high). 5 high	8.5 m	8.5 m	8.5 m	9.5 m	9.65 m	9.7 m	9.7 m	11.0 m	11.0 m	9.75 m	9.70 m	9.70 m

*Cost range/pallet stored (£) includes storage equipment only (1990 prices).

A typical layout is shown in Figure 8.18 and a 5000 tonne store equipped with powered mobile narrow aisle racking is shown in Figure 8.19.

8.5 Comparisons of racking systems

The tables on pages 244–247 compare the different types of racking systems, summarizing their physical capacities (Tables 8.1 and 8.2), relative merits in practical terms (Table 8.3) and factors of capacity and cost which affect their overall value for money (Table 8.4).

9 Mechanical handling

A.F. HARVEY

9.1 Introduction

Until the 1950s, fish landed from Dogger Bank trawlers was stacked in crates in frozen stores. This was done by hand at UK ports such as Grimsby, Lowestoft, and Great Yarmouth. The stores were not large (typically 20 m × 10 m and only 3–4 m high). An attempt was made to keep the temperature down to −25°C, but usually −20°C, at best, was achieved.

In ambient temperature food stores, electric FLTs with pallets were being widely employed. Then saving in labour was enormous. Could FLTs and pallets be used in the frozen fish stores?

The 3 m gangway necessary for FLT operation wasted too much space. On the other hand, one man driving his truck into the stores and out again was economically preferable to a squad hand-stacking crates of fish, until they too became almost frozen. The balance tipped in favour of trucks plus pallets when the *reach fork lift truck* which would operate in a 1.7 m gangway, was popularized in the mid-to-late 1950s (Figure 9.1).

Two FLTs were bought to operate in a small fish store in Lowestoft. Within a week there were serious technical problems. The nylon tyres, then used as standard, cracked off the wheels; oils and greases froze; hydraulic seals were cut, and the hydraulic pumps literally burst. The battery, too, seemed 'dead'— not that it mattered much, under the circumstances! In the short term, little could be done for the trucks themselves, other than to repair them and then to agree a strict operating procedure in which they entered the cold store merely to place or retrieve one load at a time. Thus, they remained in the cold temperature for only a short time and the temperature of the components did not fall seriously. Fortunately, the customer concerned accepted this situation, mainly becaue the truck operators also preferred the 'in-out' regime.

That was a start, but not a happy one; it left the truck manufacturer with a problem if his products were to be seriously considered for cold store handling. But problems are for solving, especially when, as in this case, the potential demand was considerable.

A development programme was launched. Various so-called 'arctic' lubricants were tried for the oil in the gear boxes, the grease in the bearings, and the

Figure 9.1 Reach FLT (*c.* 1950s).

hydraulic fluid. (Brakes, in those days, were mechanical.) After much trial and error, suitable lubricants were discovered. A mix of rubber which withstood the cold and supported the wheel loads was devised. A means of keeping water/ice out of the hydraulic system was installed. The 'electrics' were considered, and heaters fitted where necessary.

In the end, a cold store specification for an FLT was drawn up, and it worked. The only trouble was, it was commercially unacceptable because it more than doubled the truck price! The manufacturer was told, in no uncertain terms, that this was quite out of the question. So the designers had to go back to the drawing board.

Figure 9.2

First of all, a serious attempt was made to determine a 'universal' specification for a cold store truck. It had to be a reach truck (to save space); it had to be a 2-tonner to cope with loads of frozen peas and ice-cream. It needed to stack loads up to about 5.5 m, yet go through doors only 2.4 m clear. Ideally, it should be capable of clamping a load in those applications where loads were suitable for clamping. This was not too difficult (Figures 9.2 and 9.3)—but

Figure 9.3

how about making it suitable for use in cold conditions, and keeping the price right?

It was decided to adopt a different approach: to take a *standard* truck and see what went wrong, then design out the problems as they arose.

The most significant fact discovered was that, provided the trucks were kept *at work* in the cold store, the low temperature, as such, was not a serious problem at all. The very fact that the electrics are not 100% efficient means that heat is generated in batteries, cables, control gear, and motors; this keeps up the temperature of the electrical components. The hydraulic system similarly suffers from losses which heat up the oil enough to prevent problems with seals, pumps, etc.

There remained the problem that when a truck was brought out after it had worked in the cold store for some time, it immediately became covered in hoar-frost outside and in, even inside the air-space of hydraulic jacks and in the tank. As the truck warmed up this hoar-frost became dew and the whole truck ran with condensation.

This created a major problem in the hydraulic system because the water emulsified the oil and froze into ice particles as soon as the truck-entered the cold. These particles soon cut the seals in the hydraulic system and ruined them. A simple, cost-effective solution was to close the hydraulic system so that it was isolated from the outside atmosphere using the bellows, or air-sac, on the tank of protected trucks. Such trucks are used today in cold stores (Figure 9.4).

To prevent long-term rusting problems, steel parts were zinc (or otherwise) treated. One or two other 'belt-and-braces' precautions were taken, like slight modifications to seal materials and somewhat 'improved' lubricants. The total results was an on-cost over standard of not more than 20%, which was acceptable to users who then had a reliable means of handling their loads in cold stores. The only snag arose if, inadvertently, a truck was left in the cold

Figure 9.4 FLT with sealed hydraulic system.

store when it was not working. Operators then had to drag it out and repairs could be expensive: they did not make that mistake often!

9.2 Developments

Having established an acceptable specification and operating technique, which was economically satisfactory, it then remained to apply these to types

Figure 9.5

of truck other than reach FLTs. This was done and has continued to be done as new types of truck have been developed (Figures 9.5 and 9.6).

A good deal of attention was given during the 1970s to protection for truck drivers. Heavy, cumbersome, arctic clothing (Figure 9.7) made operation difficult. Big boots were not ideal for moving from accelerator to brake pedals; heavy padded gloves gave inadequate 'feel' of hydraulic levers; helmets restricted vision. Furthermore, it is known that cold affects the quick thinking or reaction time of a truck driver, who thus becomes more accident-prone. Added to that, a seated driver is not exercising his body, so the cold

Figure 9.6

Figure 9.7 FLT operator with heavy clothing.

tends to penetrate the back—the kidneys in particular—with possible serious effects.

The immediate thought was to use a sort of old-fashioned flying suit with electrically heated boots, gloves, and clothing. After all, there was plenty of energy available in the batteries on the truck: a few watts used to heat the operator's clothing would be of no consequence. Unfortunately, the demand for heated flying suits had disappeared (modern aircraft all had heated and pressurized cabins). So no supplier could be found to make the comparatively few suits required by cold store truck operators. The idea, though good, had to be abandoned. This meant that to keep a truck running, at least two operators had to take it in turns—typically half an hour in and half an hour out. This obviously increased the running costs very considerably. But the development engineers returned to the aircraft idea. If a warm working environment filled the bill there, why not devise a heated cab for a cold store truck?

A great deal of experimental work was needed. Again the philosophy of meeting problems as they arose was adopted and the willing (and patient) cooperation of cold store operators proved invaluable (Figure 9.8). The eventual outcome was entirely satisfactory. But it is worth noting some of the problems that had to be solved:

Figure 9.8 FLT operators with heavy clothing, and FLT operator in a heated cab truck.

- The cab must provide the *strength* normally afforded by the overhead guard
- The cab must not seriously impede the operator's vision
- It must have no projections (e.g. door handles) and must not increase the overall truck dimensions excessively
- The cab must not impede the operator's movements
- Its glazing must be of a material which is:
  —tough
  —a good heat insulator
  —scratch resistant
  —resistant to misting
  —shatterproof (vital to avoid contamination of products)
  —easily replaced
  —non reflective
- The cab must quickly warm up but the interanl temperature must then be maintained by thermostat, within close limits
- The trucking system must require few door openings, as this is when serious heat loss occurs
- Communication with the 'outside world' is essential as any well-insulated cab is, effectively, an acoustic shroud too

- A means is essential to maintain the temperature inside the cab when the truck is idle for long periods, to reduce battery drain
- Fresh air i.e. cab ventilation, is essential and cold internal surfaces liable to be touched by the operator must be padded
- The cab must not seriously restrict access to truck components for essential maintenance

All these problems have been overcome (Figure 9.9), although some degree of compromise has been necessary in order to achieve a result which is

Figure 9.9 Coldstore FLT fitted with heated cab.

successful but also commercially acceptable. Today, a successful cold store cab fitted to a high-lift reach FLT adds about 20% to the basic chassis price; fitted to a VNA truck adds about 10%.

In summary, after considerable development, a satisfactory, practical cold store protection is available for battery powered trucks. In addition, reach and VNA trucks can be equipped with a cab which enables one operator to work a full shift, without the encumbrance of thick clothing. This adds to the initial cost of the truck, but the return on investment is immediate because only one truck operator is needed, instead of at least two, to keep the truck running continuously.

9.3 The present day

Change in operating temperature

All the early work was done in stores where the temperature was, nominally, $-25°C$: in practice this was usually about $-20°C$. The 'standard' cold store temperature is now $-30°C$, and, in practice, this is usually maintained. Providing lubricants are carefully selected and providing the disciplines of operation are observed, this lower temperature creates no problems at all.

Change in cold store size

Trucks were originally employed in tiny stores where, previously, all loads had been manhandled. 1000 m^3 was typical. However, the very use of trucks meant that larger stores were appropriate. In fact, to achieve a reasonable amount of storage, the stores had to be larger to give space for truck operation. More significantly, the market demand for frozen food has developed. Thus cold stores have grown larger in area and also significantly higher. Reach trucks can readily stack to 9 m, giving a height over the stack about 10.5 m. VNA trucks can go to about 13 m, giving a height over stack of about 14.5 m.

A single store 60 m–70 m square and with 16 m under the ceiling is now typical: about 75000 m^3.

Improved space utilization

Obviously, the best space utilization is where stock is placed from floor to ceiling, in a sort of block stack. Working from one end of a store it can, theoretically, be filled up solid, right to the door. This could be regarded as 100% utilization of available space. But this is obviously not practical:

- Because the lower loads will crush
- Because it is impossible to access the loads first put in without removing all the others, first in becomes the last out (FILO).

Nonetheless, with certain goods e.g. ice-cream, it is sometimes reasonable to block stack, as it is termed. The only waste space is then that necessary between stacks to permit movement plus the width of a gangway at each end of the store. Thus a store 70 m long can contain a block about 64 m long (with 3 m gangways each end). In a 50 m width a maximum gap of 100 mm between adjacent loads (which are, say, 1 m wide each) gives 45 m of occupied space. If the store is, say, 7.8 m high, 5 × 1.5 m high loads can be placed giving a 7.5 m high stack. The space utilization is:

$$\frac{64}{70} \times \frac{45}{50} \times \frac{7.5}{7.8} = 0.79 \text{ or about } 80\%$$

It is sometimes practical to clamp loads, which saves the volume space usually lost to pallets (see Figure 9.3). However, it is then necessary to increase the gap between loads to accommodate the clamp arms, so the space utilization, using block stacking and clamp trucks, still rarely exceeds 80%.

To afford ready access to every load it is necessary to provide *gangways* down the store. Using reach trucks these are typically 2.5 m wide. A typical layout is shown in Figure 9.10. Whilst, in the store length, a 64/70 ratio still maintains and 7.5/7.8 still applies to the height, the width ratio obviously

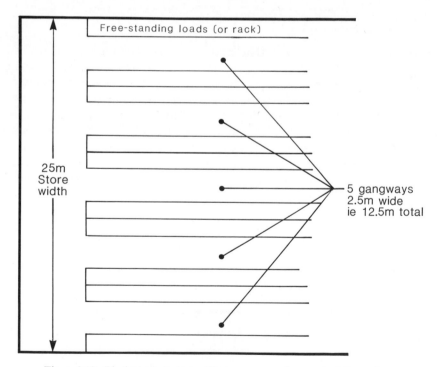

Figure 9.10 Block-stacked store with gangways to show typical dimensions.

Five racked loads

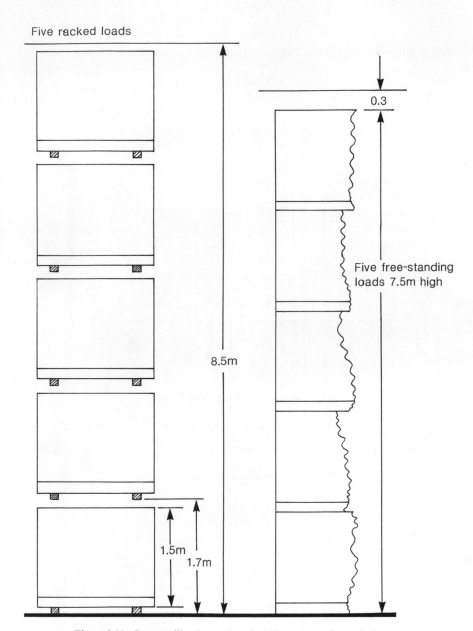

8.5m

0.3

Five free-standing
loads 7.5m high

1.5m

1.7m

Figure 9.11 Space utilization and typical dimensions of a racked store.

becomes 0.5 (Figure 9.10). Thus, the space utilization becomes:

$$\frac{64}{70} \times 0.5 \times \frac{7.5}{7.8} = 0.44 \text{ or, say, } 45\%$$

If loads are racked, the rack beams waste 100 mm height and space is needed above every load (Figure 9.11). Thus a 1.5 m high load occupies, typically,

Figure 9.12 Drive-in racking.

1.7 m height. Space utilization then becomes:

$$0.44 \times \frac{1.5}{1.7} = 0.39 \text{ or, say, } 40\%$$

This is, of course, a dramatic reduction from the 80% associated with pure block stacking but it is the price that has to be paid for:

- Access to every load
- No danger of squashed loads
- The ability to use reach FLTs

However, various attempts have been made to improve this 40% figure. For example:

- *Drive-in racking* (Figure 9.12)   This denies ready access to every load and is limited to modest heights: but it gives space utilization of about 50%
- *Mobile racking*   This allows easy access to every load but, because one gangway can serve, typically, up to 10 racks, the space utilization is around 60%
- *VNA or turret trucks*   This merits closer study because the VNA system is certainly an alternative to reach trucks and it employs standard (i.e. not drive-in and not mobile) racking.

The 70 m long store can be racked to leave, say, a 5 m wide aisle at one end only. However, pick-up and despatch (P&D) points are needed for each aisle because VNA trucks cannot fetch and carry their own loads. So a further 2 m is 'lost', leaving a racking length of 63 m. Across the VNA store aisles are typically 1.6 m wide, compared to 2.5 m for reach trucks (Figure 9.13). In height, this is a racked store so the 1.5:1.7 ratio applies.

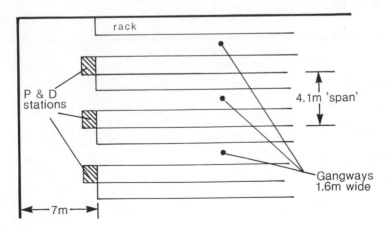

Figure 9.13   Space utilization in a racked store using VNA trucks.

Hence the space utilization of a VNA system is:

$$\frac{63}{70} \times \frac{2.5}{4.1} \times \frac{1.5}{1.7} = 0.48 \text{ or, say, } 50\%$$

Access to every load is available, with the additional advantages of the possibility to stack to about 13 m and a higher rate of work throughput, compared to reach FLTs. However, it must be stressed that VNA trucks demand system discipline:

- Floors must be smooth and level
- Load sizes must be carefully controlled
- A guide system is necessary in the aisles.

The advantages and disadvantages of different stacking systems are compared in Table 9.1.

Operating techniques

The present-day operating temperature is −30°C. Cold store size has increased, and space utilization is, as always, important because it costs money to keep every cubic metre cold.

Speed of operation—getting loads in or out rapidly—is an important economic consideration. Even more important is the need to enable one operator to endure a full shift in reasonable comfort. Hence, the importance of a properly designed, heated cab on FLTs.

But an even more significant change has been brought about recently because:

- The taller reach trucks require high doors to enable them to enter or leave the cold store.
- VNA trucks are too tall anyway.

Table 9.1 Space utilization: advantages and disadvantages of stacking systems.

System	Space factor	Stacking	Advantages	Disadvantages
Block stack	80%	'Free'	Highest factor	Limited access, slow
Reach FLT	45%	'Free' but in gangways	Access to loads	Load, squash
Reach FLT	40%	Racked	Full load access	Poorest factor
Reach FLT	50%	Drive-in racked	Better factor	Slow, poor access
Reach FLT	60%	Mobile racking	Better factor	—
VNA (turret)	50%	Racked	Store height, full access, good work throughput	Specialized

Thus, the fundamental rule of the 'in–out' operation (i.e. take the truck out when it is not at work) just cannot be applied.

Once again the application engineers of the truck manufacturers had to do some thinking. With hindsight, the operating technique which has evolved has much to commend it, over and above the 'in–out' routine. It is a 'stay-in' technique, in which the truck certainly stays in the cold store during all normal periods of operation and is brought out, at most, say, once a month, for maintenance only.

This means that loads must be fed into the store and removed from it either by simple, small, transporter trucks or by conveyors. This, in turn, means that the input/output doors of the store need to be of only modest size, thus reducing 'cold loss' considerably. (Of course, one set of doors is necessary to allow the trucks to be removed for maintenance but they are used only once per month, so the loss incurred is minimal.)

On any 'stay-in' operation, condensation is no longer a problem but pure low temperature is. Whilst the cold store specification of trucks can cope with this, certain operating disciplines are essential. These are:

- The batteries must be changed every eight hours, regardless of use. (The batteries are charged at normal ambient temperatures anyway.)
- If a truck is left idle for any period longer than a quarter of an hour, its cab heater must be plugged into an AC mains supply in order to maintain the temperature of the cab, without battery drain
- When the truck returns to the cold store after maintenance it must be thoroughly dry. If, rarely, this creates a problem it is usual to apply pressurized warm air jets, powered from the waste heat extracted from the cold store

Summarizing, therefore, the modern cold store FLT:

- Has a heated cab
- 'Stays in', but has its battery charged every working shift
- Is 'fed' by transporter trucks or conveyors
- Operates the simple 'stay in' discipline as above

The main advantages over the 'in-out' routine are:

- Minimum cold loss from the store
- One operator per truck per working shift
- A simple truck specification or reduced maintenance because condensation is not a problem
- Faster, more efficient, operation

Further observations

Taking a truck into a cold store and using it there adds to the load on the refrigeration plant. By how much? Firstly, there is the mass of the truck itself which has to have its temperature reduced to $-30°C$. The calculation usually made assumes the whole truck to be made of steel, with specific heat 0.115. Secondly, some of the battery energy is lost as heat when the truck operates. The calculation made assumes that 50% of the energy has to be 'overcome' by the refrigeration plant during each working shift.

In certain cold store operations an order-picking duty has to be performed. This is essentially labour-intensive and every effort should be made to avoid it, e.g. by insisting that retail outlets take *full* loads. This is certainly not always practicable, particularly on slow-moving goods (like seafood, confectionery, etc.) So, like it or like it not, some order-picking in cold store is unavoidable. The sole recommendation must then be to ensure that all the 'right' order-picking techniques, as applied to normal ambient operations, are rigorously applied in the cold store context.

Load discipline is a feature of good materials handling anyway: i.e. no broken pallets; no broken cartons; no protruding cartons. Such discipline becomes, if anything, even more important in cold stores. This is especially true if VNA trucks are used.

Trucks run on wheels; wheels run on floors. Although ice build-up is not a feature of the floor of a modern cold store (it used to be a major hazard!) it should be remembered that an uneven floor does not assist good trucking. A slippery floor can be lethal. (A few tonnes of truck and load sliding out of control is a major danger to men and goods.)

9.4 The future

Thirty years after FLTs were first used in cold stores, they have become the most popular form of materials handling in cold stores today. There is nothing to suggest that FLTs will not continue to do the major handling work in the cold stores of tomorrow. But will there be further changes?

What about even lower temperatures for example? Japanese truck manu-facturers already offer trucks 'suitable for $-40°C$', although this claim has yet to be proved in practice, in Europe at any rate. In fact, there could be major problems because there is a dramatic change in the ductility of constructional mild steel at about $-35°C$. So 'special' steel might be necessary for the whole truck chassis; and, indeed, even for the racking which supports loads. Such 'special' steel would be costly. So, lower temperatures, which might provide longer food preservation, would mean a fundamental rethink for truck/rack manufacturers. In any case, the heat insulation of the store and the necessary energy to maintain lower temperatures might also need a rethink. It would seem that $-30°C$ will be the standard for some time yet.

More positively, however, the VNA truck is gaining popularity because it enables a better space utilization (50% as opposed to 40% for a reach truck); it enables higher stacking (13 m instead of 9 m to top shelf); and it is capable of a higher work throughput (up to 35 loads per hour, instead of about 25 for a reach truck).

Crane storage has also been used when stacking in excess of 10 m in height. Indeed automated (driverless) storage has been tried. However, any such system is inherently inflexible and is therefore restricted to a store which is unlikely to change in any aspect for many years.

Although VNA trucks demand discipline they do retain most of the flexibility inherent to industrial trucks. Understandably, cold store users are considering them more and more.

The final point, inevitably, is one of cost. All stores and warehouse handling adds to cost. Cold store space/handling is inevitably more costly than normal. The VNA system can often be the most cost-effective, especially in larger cold stores. Although the reach truck will continue in popular demand by cold store users, the use of VNA equipment will continue to grow.

10 Cold store doors

W. van der SPEK and G.S. ATKINSON

Symbols and units

b	Width of cold store doorway	m
g	Acceleration due to gravity; normally taken as 9.81	m/s^2
h	air enthalpy	h
p_{sat}	Saturated vapour pressure of water vapour	mbar
p_s	Actual vapour pressure of water vapour	mbar
p_{atm}	Atmospheric pressure	mbar
p_{act}	Actual air pressure $(p_{atm} - p_s)$	mbar
R	Characteristic gas constant; for air, R = 287.1	J/kg K
s	Ratio of ambient air density to cold store air density, allowing for moisture content	
T	Temperature	°C
t	Door-open time	s
t	Periodic time of door-open cycle	s
v	Specific volume of dry air	m^3/kg
w	Mass of water vapour associated with 1 kg of (dry) air, called the specific humidity	kg moisture/ kg dry air
$\mathring{V}$	Rate of air movement through open doorway	m^3/s
y	Height of cold store doorway	m
ϕ	Relative humidity $= p_s/p_{sat}$	
ρ	Air density including moisture content	kg/m^3
γ	Kinematic viscosity	m^2/s
Φ	Correction factor to Tamm's equation to allow for short door-open times	

10.1 Introduction

With increasing demand for cold storage space, the volumes of individual stores have increased tremendously. The price of land has become a major cost in cold store construction, so it seems sensible to save ground space and build cold stores with high ceilings. For a long time, the heights of cold stores were limited to 2–3 m by the maximum lift of goods handling equipment such as the fork lift truck. The recent development of high lift trucks has meant that modern cold stores can now be built with ceiling heights of over 10 m.

The efficient operation of large modern cold stores has necessitated complete mechanization, and the most important piece of goods handling equipment has been the fork lift truck. The use of this vehicle has made the large, hinged type of cold store door largely obsolete. Traffic densities tend to be high, especially during harvest time, with doors opening and closing every few minutes, and the increased sizes of loads requires very large door apertures. For truck drivers to dismount in order to open and close doors would be very inefficient, involving loss of time and productivity and considerable heat gain through open doorways. Power-operated sliding doors were therefore developed which could be opened and closed quickly and automatically (Figure 10.1).

In this chapter, the various types of power-operated sliding doors and their ancillary fittings will be described, together with discussion of their applications, advantages and disadvantages. In cold stores, careful consideration must be given to the following points:

- Size of the door aperture
- Selection of door type
- Door-opening controls
- Door protection and safety
- Door maintenance

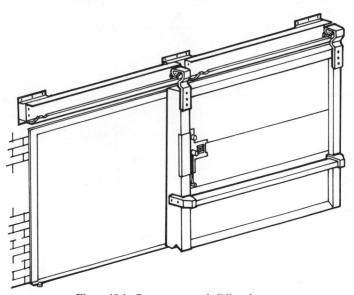

Figure 10.1 Power operated sliding doors.

10.2 The effects of air leakage

Ambient air is less dense than cold store air and is therefore relatively buoyant when the two are mixed. When a cold store door is opened, cold air spills out along the floor like water through a sluice, while warm air enters above (Figure 10.2). The warm air initially rises towards the cold store ceiling, but gradually it mixes with the cold store air to form one homogeneous mass. During the mixing process a great deal of heat and moisture is transferred. Infiltrating air adds to the heat and moisture loads imposed on the store, as well as to the store operating expenses.

The ambient air has a much higher moisture content than the cold store air, especially in summer. In cooling down to store temperatures, infiltrating air must shed moisture as its humidity and temperature approach those of the store. Much of the excess moisture is shed as ice on surfaces close to the door, and on the evaporator coils. Ice represents a serious problem if it is allowed to form on the door and door frame: it would interfere with the door seals and eventually clog the operating mechanism. The cold store door or its frame must incorporate a low voltage heating system to prevent the accumulation of ice on or near to its sliding surfaces and seals. Moisture shed onto the evaporators adds to defrosting loads, increasing the amount of energy required for defrosting and the frequency and length of the defrosting cycle. Calculation of the moisture gain per unit mass or per unit volume of infiltrating air can be made, provided that the temperatures and relative

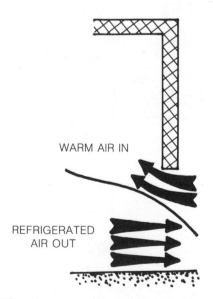

WARM AIR IN

REFRIGERATED
AIR OUT

Figure 10.2 Air movement through an open coldstore door.

humidities of the ambient air and cold store air are known:

$$\log p_{sat} = \frac{7.5T}{(237.3 + T)} + 0.78571 \qquad (10.1)$$

$$\log p_{sat} = \frac{7.5T}{(265.5 + T)} + 0.78571 \qquad (10.2)$$

$$p_s = \phi \cdot p_{sat} \qquad (10.3)$$

$$w = \frac{0.622\,p}{(p_{atm} - p_s)} \qquad (10.4)$$

$$v = \frac{287.1 \times (T + 273.15)}{(p_{atm} - p_s) \times 100} \qquad (10.5)$$

Equations 10.1 and 10.2 are the Magnus formulae and are used to calculate the saturated vapour pressure of moist air. Equation 10.1 gives the SVP of ambient air, and Equation 10.2 gives that of cold store air. The air temperature (T) is in °C, and pressure (p) is in millibars. Their use can be illustrated best by a worked example.

Example 1   Ambient air at 25°C and 60% relative humidity infiltrates a cold store of temperature -30°C in which the relative humidity is 95%. Estimate the moisture load on the store, per unit volume of infiltrating air. The atmospheric pressure is 1013.25 mbar.

Ambient air:

$$\log p_{sat} = \frac{7.5 \times 25}{(237.3 + 25)} + 0.78571$$

$$= 1.50054$$

$$p_{sat} = 31.66\,\text{mbar}$$

$$p_s = 0.60 \times 31.66$$

$$= 19.00\,\text{mbar}$$

$$w = \frac{0.622 \times 19.00}{(1013.25 - 19.00)} = 0.01188\,\text{kg/kg dry air}$$

Cold store air:

$$\log p_{sat} = \frac{7.5 \times (-30)}{(265.5 - 30)} + 0.78571$$

$$= -0.1697$$

$$p_{sat} = 0.68\,\text{mbar}$$

$$p_s = 0.95 \times 0.68$$

$$= 0.64\,\text{mbar}$$

$$w = \frac{0.622 \times 0.64}{(1013.25 - 0.36)} \qquad = 0.00039\,\text{kg/kg dry air}$$

∴ Every kilogramme of air (not including the mass of its moisture content) sheds $0.01188 - 0.00039$ kg of moisture $= 0.01149$ kg of moisture.

The specific volume of ambient air

$$v = \frac{287.1(273.15 + 25)}{(1013.25 - 19.00) \times 100} = 0.861\,\text{m}^3/\text{kg}$$

Therefore the moisture load per unit volume of infiltrating air $= \dfrac{0.01149}{0.861} = 0.01335\,\text{kg/m}^3$

Example 1 shows that a seemingly modest amount of ice will be deposited in the cold store for each cubic metre of infiltrating air. Modern cold store doors are often large in order to accommodate the large trucks and loads which must pass through them, and the doors of a busy store are opened frequently during the working day. So the actual volume of infiltrating air will be very great and this makes the total amount of ice deposited considerable.

The heat load imposed on the store by air infiltration is part 'sensible' due to the cooling of air and water vapour, and part 'latent' due to the condensation of excess moisture. Equation 10.6, although simple, enables air enthalpy to be calculated to an acceptable level of precision.

$$h = 1.005T + w(2500 + 1.86T) \qquad\qquad (10.6)$$

The datum temperature for Equation 10.6 is 0°C, so for temperatures below freezing the enthalpy of air may well turn out to be negative. This presents no difficulty, since it is changes in enthalpy which are being considered, not absolute values. This is illustrated in Example 2.

Example 2 Given the ambient and internal cold store conditions of Example 1, calculate the heat load imposed on the store per unit volume of infiltrating air.

Ambient air:

Taking the value of w for ambient air calculated in Example 1,

$$h = (1.005 \times 25) + 0.01188[2500 + (1.86 \times 25)]$$

$$= 55.38\,\text{kJ/kg}$$

Store air:

$$h = 1.005 \times (-30) + 0.00039[2500 + (1.86 \times -30)]$$
$$= -29.20 \, \text{kJ/kg}$$

Heat load on store $= 55.38 - (-29.20)$

$$= 84.58 \, \text{kJ/kg air}$$

From Example 1, the specific volume of the ambient air $= 0.861 \, \text{m}^3/\text{kg}$, therefore heat load on store

$$= \frac{84.58}{0.861} = 98.23 \, \text{kJ/m}^3$$

So far the store moisture and heat loads imposed by one cubic metre of infiltrating air have been calculated. When the total loads on a store are required, then the most difficult problem is to estimate the total volume of infiltrating air. This is affected not only by temperature, pressure and humidity of the ambient air and cold store air, but also by the width and height of the door aperture, the frequency and durations of openings, the type of traffic, and even the sizes of goods stored and their storage patterns. It is impossible to make absolutely precise assessments without doing lengthy, expensive, and probably inconvenient tests on individual stores. Various investigations have been carried out to determine the rates of cold air loss through open cold store doorways [8, 14, 31, 32, 33].

As part of his research, Tamm [33] developed a useful and concise equation to calculate the rate of loss of cold air through an open cold store doorway:

$$\mathring{V} = \frac{2 \, b \, y}{3} \sqrt{\frac{2 \, g \, y(1-s)}{(1+s^{1/3})^3}} \tag{10.7}$$

Pham and Oliver [14, 31] developed tracer gas techniques using sulphur hexafluoride (SF_6) to measure the rate of air flow through open doorways. They carried out tests on stores ranging in volume from $177 \, \text{m}^3$ to $37\,000 \, \text{m}^3$ with apertures ranging from $1.08 \times 1.98 \, \text{m}$ to $3.0 \times 3.6 \, \text{m}$. It appears that the flow tests were made on doorways which were permanently open. They concluded that Tamm's theoretical equation was valid, but that it should be modified by including a discharge coefficient of 0.68 ± 0.04. Applying their modification Tamm's equation (10.7) becomes:

$$\mathring{V} = 0.453 \, b \, y \sqrt{\frac{2 \, g \, y(1-s)}{(1+s^{1/3})^3}}$$

Gosney and Olama [8] conducted tests on cold store models, simulating the higher density of the cold store air by mixing the air inside their models with carbon dioxide gas. Their work tended to confirm Tamm's equation and the work of Pham and Oliver, but concluded that the coefficient should be 0.441 rather than 0.453. However, allowing for the fact that the Pham and Oliver tests were made on full size stores with refrigerated air, it seems reasonable to accept the higher coefficient of Pham and Oliver. A further conclusion of both research teams was that under steady conditions, the mass flow of cold air out of a cold store must be exactly equal to the mass flow of ambient air into it.

Equation 10.7 applies only to constantly open doorways. In practice working doors are open only for short lengths of time at fairly frequent intervals. Hence, estimation of the rate of cold air loss through working doors requires further modifications to Tamm's formula. It is recognized by various authorities that during the first few seconds of door-open time the air flow accelerates. After about 7 seconds it reaches a steady maximum flow rate. Thus, if the door-open time is short, Equation 10.7 is likely to overestimate the flow rate. Gosney and Olama included a further factor Φ in Tamm's equation which compensates for short door-open times:

$$\Phi = \left[1 - \frac{0.0112}{(\gamma_o \cdot t_o^{0.3565}/y)} \right] \left(\frac{t_o}{t_c} \right)^{0.1} \tag{10.8}$$

This factor Φ modifies Tamm's equation yet further to the form:

$$\mathring{V} = 0.453 \, by \sqrt{\frac{2 \, g \, y(1-s)}{(1+s^{1/3})^3}} \, \Phi \tag{10.9}$$

The kinematic viscosity of air may be conveniently calculated from the formula

$$\gamma_o = 1.33 \times \left(\frac{273 + T}{273} \right)^{1.75} \times 10^{-5} \, \text{m}^2/\text{s} \tag{10.10}$$

Example 3   The air within a cold store is at a temperature of $-30°C$ with 95% relative humidity. The ambient air is at 25°C with 60% relative humidity. The cold store door is 2.0 m wide and 2.5 m high. It is open for an average time of 30s with a frequency of 30 times per hour, for 12 hours per day, during 340 days per year. The coefficient of performance of the store's refrigerating equipment is 1.90. The atmospheric pressure is 1013.25 mbar.

(i) Estimate the cost of air infiltration through the door if the cost of electric power is 4.20 pence per kWh.
(ii) Estimate the moisture load on the store.

(In this example suffix 'o' refers to ambient air and suffix 'i' to cold store air.) Making use of the quantities calculated in Examples 1 and 2, the density of the ambient air including the mass of its moisture content

$$\rho_o = \frac{1 + 0.01188}{0.861} \qquad\qquad = 1.1752 \, \text{kg/m}^3$$

The specific volume of air in cold store

$$v_i = \frac{287.1 \times (273.15 - 30)}{(1013.25 - 0.64) \times 100} = 0.689 \, \text{m}^3/\text{kg}$$

Therefore density of cold store air

$$\rho_i = \frac{1 + 0.0039}{0.689} = 1.4519 \, \text{kg/m}^3$$

and

$$s = \frac{1.1752}{1.4519} = 0.8094$$

$$\gamma_o = 1.33 \times \left(\frac{273 - 30}{273}\right)^{1.75} \times 10^{-5} = 1.085 \times 10^{-5} \, \text{m}^3/\text{s}$$

$$\Phi = \left[1 - \frac{0.0112}{(1.085 \times 10^{-5} \times 30)/2.5^{0.3565}}\right]\left(\frac{30}{120}\right)^{0.1} = 0.6339$$

$$\mathring{V} = 0.453 \times 2.0 \times 2.5 \times \sqrt{\frac{2 \times 9.81 \times 2.5 \times (1 - 0.8094)}{(1 + 0.8098^{1/3})^3}} \times 0.5424$$

$$\mathring{V} = 1.6329 \, \text{m}^3/\text{s}$$

The mass flow of air out of the store, discounting the mass of its moisture content

$$= \mathring{V}/v$$
$$= 1.6329/0.689 \qquad = 2.370 \, \text{kg/s}$$

This will be the same as the mass flow of (dry) air into the store, so

Heat load $= 2.370 \times [55.38 - (1 - 29.20)] = 200.4 \, \text{kW}$

But the actual door-open time

$$= \frac{30 \times 30 \times 12}{60} \qquad = 180 \, \text{min/day}$$

So average heat load spread over entire day

$$= 200.4 \times \frac{180}{24 \times 60} \qquad = 25.1 \, \text{kW}$$

Average power consumption required as a result of air

infiltration $= 25.1/1.90 \qquad = 13.2 \, \text{kW}$

(i) If cost of electric power i.e. 4.20 p/kWh,

Annual cost $= 13.2 \times 24 \times 0.042 \times 340 \qquad = £4\,520$

(ii) Mass of infiltration air, discounting mass of moisture content

$$= 2.370 \times 30 \times 30 \times 12 \qquad = 25\,596\,\text{kg/day}$$

Mass of condensing moisture, i.e. moisture load

$$= 25\,596 \times (0.01188 - 0.00039) \qquad \underline{294.1\,\text{kg/day}}$$

10.3 External cold store doors

Modern cold store doors are of sandwich construction, similar to cold store walls. Thick insulation is sandwiched between protective and supporting laminae, and bonded to them to form rigid panels. The panels are mounted in strong frames and the insulation may be bonded to these as well. Modern door insulation usually consists of expanded polystyrene or polyurethane foam. The trend is to construct larger doors with removable modular panels which, if damaged, can be replaced quickly and economically. Laminae are of various materials including stainless steel, galvanized steel, duralumin, fibreglass or an acrylic. Frame materials can include steel, duralumin, and kiln-dried timber. Surface finishes should be tough, hygienic, and washable. Various finishes are available including PVC, stainless steel and baked enamel.

Door panels should be impervious to moisture. Porous panels would permit water vapour to migrate through the door from the warm exterior towards the colder interior. The vapour would condense and freeze as it neared the inside panel, and the layer of ice formed would gradually increase in thickness until it pervaded the whole door. If the process were allowed to continue, frosting would eventually be observed on the outer surface of the door. Not only would the door increase considerably in weight, it would also have no effective insulation. Metal laminae form impervious moisture barriers if they are continuous and all joints are carefully sealed. The usual material for sealing is white, silicon-based mastic.

The recommended thickness of insulation for external doors depends on the store temperature. Wright [25] bases recommended insulation thicknesses on the maximum differences between store and ambient temperature, whilst several major companies have made recommendations similar to those given in Table 10.1 for thicknesses of polyurethane foam insulation based on store temperature.

Since modern cold store walls of sandwich construction cannot support heavy loads, the doors are usually supported by strong, precision-built metal frames (Figure 10.3). These frames should be able to transmit the loads incurred to the cold store floors and be able to carry the weight of the doors and their actuating mechanisms. The joints between door frames and cold store walls should be carefully sealed with mastic to prevent the ingress of water vapour.

In order to ensure a flat and level surface, the threshold of the door should be

Table 10.1 Insulation recommendations based on store–ambient temperature difference.

Maximum temperature difference (°C)	Insulation thickness (mm)
28	51
28–42	76
42–56	102
56–83	152

Table 10.2 Insulation recommendations based on store temperature.

Type of store	Store temperature (°C)	Insulation thickness (mm)
Cool areas		51
Coolers	2–8	76
Coolers	above 0	102
Freezers	down to -23	102
Freezers	-23 to -54	152

fitted with a hard wearing metal sill to give a surface onto which the door can seal. This sill should be fitted in place before the floor finish is laid, in order to give a datum for the finishing screed. In the case of very low temperature applications the sill should incorporate a heating element.

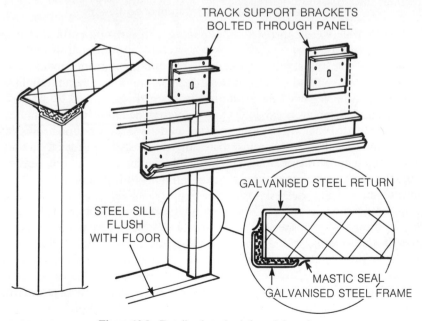

Figure 10.3 Details of steel-reinforced door frame.

Wherever possible, doors should be mounted on the outsides of cold stores as the harsh conditions inside can greatly interfere with bearings and actuating mechanisms. Also, doors fitted on the insides are likely to receive less attention and maintenance simply because it is uncomfortable for maintenance mechanics to work in low temperatures.

The principal purposes of cold store doors are to prevent the infiltration of warm air and the escape of cold air, and to permit the passage of traffic. These two functions are conflicting: an open door cannot prevent infiltration and a closed door cannot permit transit. However, in order to minimize the infiltration of air the doorway should be made as small as possible and the door-open time as short as possible.

Energy loss cannot be the main factor in sizing cold store doorways, but rather the size of the largest load which has to be admitted. Modern stores are built to heights often exceeding 10 m and goods are stacked to within less than a metre of their ceilings. In order to stack to such great heights high-lift FLTs which have been developed in recent years are used. These form the bulk of the traffic which must pass through a door, and expensive collisions between trucks and doors are frequent, so reasonable width allowances must be made in sizing doorways. FLT drivers seem to be able to estimate the widths of their loads accurately, but experience shows them to have difficulty in estimating load heights, so generous height allowances should also be made. The following formulae are offered:

$$\text{Minimum } b = 0.8 \text{ m} + \text{maximum load width}$$
$$\text{Minimum } y = 0.3 \text{ m} + \text{maximum load height}$$

If trucks are obliged to approach doors obliquely then the estimate of width b should be revised upwards.

Much time will be lost and a great deal of air infiltration will occur if truck drivers have to open and close cold store doors by hand. For minimum door-open times to be achieved doors should be opened and closed by powered mechanisms which drivers can operate without having to dismount from their trucks. Pull-cord actuators are widely used to initiate door opening (Figure 10.4). These usually require truck drivers to stop in order to actuate door-opening mechanisms, but can be sited so that they do not have to dismount. This goes some way to preventing collision damage, without causing unnecessary delays. Door-opening speeds have been quoted in various references and some recommendations are given in Table 10.3. Clearly, door-opening speed calculations will need to take into account different door widths, so the table should be treated as only a guideline.

Wilder recommends that good practice would require FLTs to be limited to speeds of 2.4–4 km/h. This would help to cut down the number of collisions and minimize the effects of those that happened, but is not commensurate with efficient goods handling. In practice truck speeds in the region of 8 km/h, a little faster than a brisk walking pace, are more common. The door-open

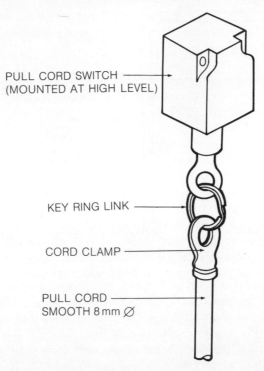

PULL CORD SWITCH
(MOUNTED AT HIGH LEVEL)

KEY RING LINK

CORD CLAMP

PULL CORD
SMOOTH 8 mm ∅

Figure 10.4 Pull-cord actuator.

speeds in Table 10.3 are based on the time required for the truck to reach the door from the pull-cord position. Careful siting of the pull-cord fixes the position of the truck when the door begins to open. Table 10.3 assumes that once the cord has been pulled the driver immediately sets off towards the door, which will be fully open by the time he reaches it. In fact drivers can often operate the pull-cord without stopping. The danger in this procedure is that collision will probably occur if door-opening is delayed in any way. It is the author's firm belief that the correct pull-cord position should place the truck as close as possible to the door. This obliges the driver to stop and wait until the door is fully open before moving and is the best way to prevent expensive

Table 10.3 Recommendations for door-opening speeds. From [4] and [18].

Truck speed	Door-opening speed	Truck-to-door distance
3.2 km/h	0.6 m/s to 0.76 m/s	
8.0 km/h	0.3 m/s	11 m
8.0 km/h	0.5 m/s	6 m

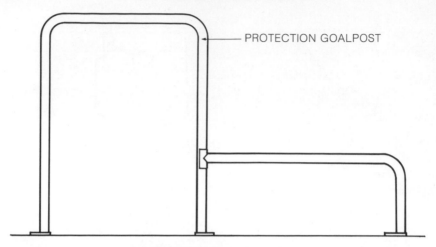

PROTECTION GOALPOST

Figure 10.5 Protection post.

collisions. If, even after this procedure, a collision should occur the truck would not have acquired any appreciable momentum before impact.

It is important to protect the door and frame by suitably fitted protection posts (Figure 10.5). These posts should be fixed securely to the floor and not connected to either the door frame or the cold store wall. They should be removable to allow repairs to be carried out to the door. It is recommended that the posts should be brightly painted, preferably in black and yellow stripes, to make them highly visible.

The siting of such posts is critical, and the centre line of the posts should line up with the edges of the door frame (Figure 10.6a). All too often this is overlooked and posts are positioned incorrectly so that they do not give full protection (Figure 10.6b). The door panel can be protected against the impact of a fully laden truck by means of a bumper bar.

If doors are constructed from a series of modular panels these can be replaced quickly and effectively in the unlikely event of damage. If doors are constructed from large, single panels then impact could cause a great deal of distortion and require difficult and expensive repairs, and prolonged and inconvenient loss of access. Correctly installed protection posts form a necessary part of a well designed door opening.

A useful service to warn drivers of overhigh loads would be a portal frame with a light, yellow-and-black striped board hung flexibly below it on light chains. The lower edge of the board should be adjusted to the maximum permissible load height. If a truck with an overhigh load passes under it the load will rattle the board and warn the driver. Such devices have long been in use at the entrances to multi-level car parks to warn the drivers of overhigh vehicles. Of course the device can only be used where there is sufficient space on door approaches.

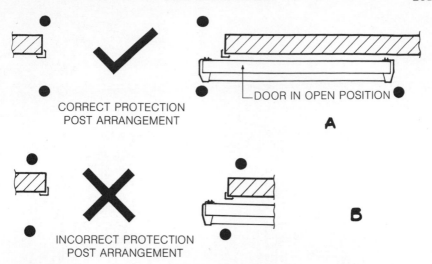

Figure 10.6 Positioning protection posts.

Since the invention by Jan Markus of the 'hermetically sealed sliding door by gravity', cold store doors have become extremely efficient.

The 'gravity sealing system' achieves opening or closing in one very simple movement with negligible friction. The door is disengaged from the frame and sill at the commencement of the opening movement. The system consists of a steel track, 38 mm square, suspended above the door at an angle of 45° to the wall and the floor. Thus, the track has two 'runways' at 90° to each other. The door is suspended from two roller-brackets, one at each end. Each bracket has two rollers which run smoothly on the tops of the two runways (Figure 10.7).

The hermetic seal is achieved very simply. At the outer limit of each runway is an indentation, into which the front roller of each bracket drops at the end of the door-closing movement. When the front rollers ride down into these indentations, the rear rollers slide laterally down their runways towards the wall and floor at an angle of 45° (Figure 10.7). The descending weight of the door presses its seals against the door frame and the sill. At the bottom of the door is a guide channel, with a cross section in the form of an inverted 'V'. A fraction outside the door aperture two cone shaped nylon cams are fixed to the floor. These cams, in conjunction with the guide channel, ensure that the bottom of the door is pressed towards the door frame as well. Gaskets fixed to the top and sides of the inner door face, and a gasket underneath the door form a complete seal. A lever is used to open the door, and its first movement takes the rollers out of the indentations, lifting the door and moving it away from the wall at an angle of 45°. Thus the pressure on the face on bottom gaskets is relieved and the door can slide without them being rubbed.

It is clear that it is a very simple matter to automate this system. As the hermetic seal and the disengagement of the door depend only on the rail

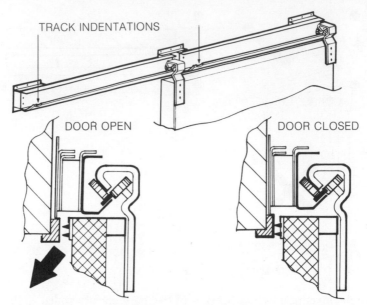

Figure 10.7   Gravity sealing system: heremetic seal mechanism.

construction, automation requires simply a means of moving the door along the tracks. Because of the smooth motion of the door, a double acting air cylinder with a stroke of only 200 mm can be used to move it (Figure 10.8). The cylinder gives an initial impulse to the door, and its momentum carries it on to complete the opening or closing movement. The door can be stopped at any instant by people or goods, thus eliminating accidents and the necessity for expensive and very vulnerable safety edges.

Double horizontal sliding doors (Figure 10.9) require space equal to a little more than half the width of the doorway to either side. It seems logical to suppose that each panel should be able to move at the same speed as the equivalent single panel, but the panels have only half the length of the equivalent single door and less stability as a result. If they are opened too quickly they tend to swing laterally in what is known as a pendulum effect, so safe opening speeds are only about 1.14 m/s.

Double bi-folding doors (Figure 10.10) are fast opening and used mainly in corridors, vestibules and other situations where lateral space and ceiling heights are restricted. The door panels are hinged together in pairs, with a pair sliding and folding to each side. As it opens its inside edges move in a straight line, guided by an overhead track, whilst the panel pairs fold together. This type of door requries more gaskets than the horizontal sliding types. Usually the upper and lower edges are sealed by sweeper gaskets which rub constantly and tend to wear more rapidly than the face gaskets used by sliding doors. Heater cables are a serious problem with this type of door because each panel

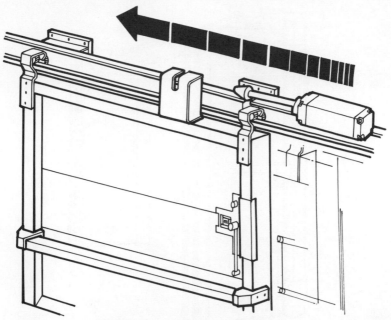

Figure 10.8 Door movement mechanism.

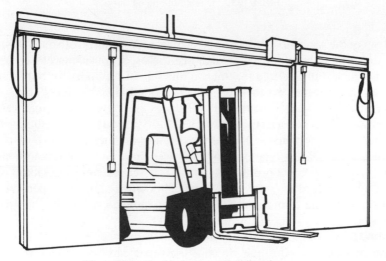

Figure 10.9 Double horizontal sliding doors.

needs to be connected to the supply. The door becomes festooned with vulnerable cables which could easily be caught up by a passing truck. Should a truck collide with the edge of a folded door the shock of the collision would be transmitted through the panels and the hinges to the door frame, and the whole assembly could be carried away. For this reason the door should either

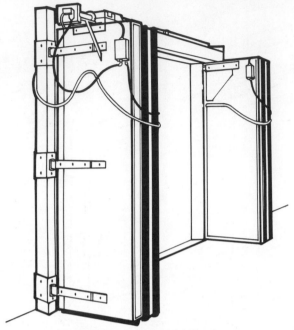

Figure 10.10   Double bi-folding doors.

fold out of the way into a recess, or be protected from such collisions by strong bollards. Single bi-folding doors can also be used, consisting of two hinged panels which slide and fold together to one side of the doorway, but these are not recommended unless there is no space available for a single sliding panel.

Where lateral space is restricted and stores are high, single-panel, vertical, sliding doors can be used (Figure 10.11). This type of door can be balanced by counterbalance weights or lifted by hydraulic or pneumatic cylinders. The opening speed of this type of door is slower than that of horizontal sliding doors, which may cause problems with FLT drivers catching their loads on the lower edges before they are fully open. They are also more expensive than horizontal sliding doors and are not recommended unless lateral space is restricted.

Various specialized doors are available on the market. Figure 10.12 shows a conveyor pass door through which a conveyor can be introduced to carry goods and materials into the cold store. They frequently form the last stage in a food-processing production line whilst the conveyor is in use. Readers are advised to confer with specialists on how best to meet their needs in this application. A trip-switch can be incorporated to link door opening and closing with operation of the conveyor.

Figure 10.13 shows a rail passing through a door opening, usually used for conveying animal carcasses into chill rooms. There are two ways of

Figure 10.11 Single panel, vertical sliding doors.

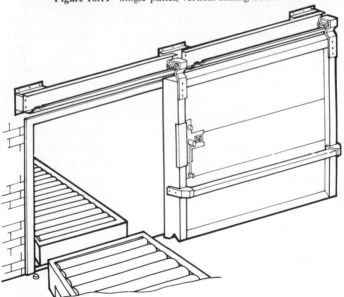

Figure 10.12 Conveyor pass.

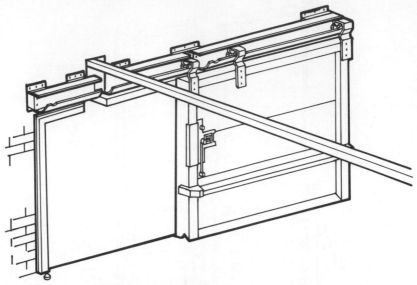

Figure 10.13 Trip-switch activated door opening.

overcoming the problem of rail penetration. The first is by having a continuous rail passing above, and breaking the track of the sliding door to allow the hooks of meat rails to pass through the opening. The gaps should be kept to a minimum and the door provided with additional bearings to enable the break to be bridged. A 'top hat' section is normally formed in the door around the meat rail, and this area should be fitted with flexible flaps. Alternatively, the door can be provided with a rail which passes underneath the track. A hinged section is swung out of the way when the door is closed, and as the door is opened the hinged section drops into position to close the gap in the rail. Various systems are available. Some are provided with mechanical devices to operate the hinged section, and others connect the section to the door and operate it by a combination of door movement and gravity action. These methods are not as safe as having a continuous, unbroken meat rail, because there is a risk that the connection will not be made correctly, and a heavy carcass could fall, causing injury to personnel.

One of the more serious effects of air infiltration is the accumulation of ice from vapour condensing on door frames. Ice accumulation on door frames prevents gaskets from sealing effectively and promotes yet more infiltration and build-up of ice. If ice is allowed to accumulate in this way it will soon prevent a door from functioning altogether.

To prevent this from happening the door, or door frame, should be fitted with heater cables or tapes. It is recommended that these should be fitted to the frame rather than to the door panel, as in the latter case it would be necessary to supply the heater power through vulnerable, trailing leads. Heater power

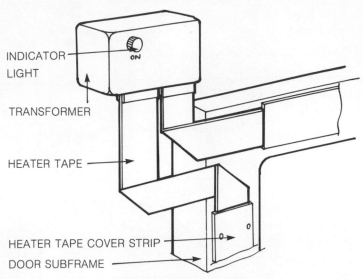

INDICATOR
LIGHT

TRANSFORMER

HEATER TAPE

HEATER TAPE COVER STRIP

DOOR SUBFRAME

Figure 10.14   Heater tape unit.

should be low voltage and low power enabling door or frame temperatures to be kept a little above freezing. For safety reasons, power is usually supplied to heaters through low voltage transformers. Heating elements are made from stainless steel wire insulated by PVC or polyester or are wide, flat, stainless steel strips covered by PVC Figure 10.14. It is recommended that these should be protected from impact damage by aluminium or steel strips and that the tapes should not come in contact with any combustible material.

Doors should have continuous gaskets to enable them to seal perfectly against their sills and frame members. Several types of gasket are illustrated in Figures 10.15–10.17.

Normally for chill room application a single gasket is used (Figures 10.15 and 10.16) whereas for low temperature applications it is better to use a double gasket (Figure 10.17) which allows heat to build-up between its two strips, keeping the immediate area free from ice. Gaskets should be reasonably flexible but at the same time tough and durable.

Figure 10.17 shows a face gasket consisting of two parallel rubber strips, used on the horizontal and vertical sliding panel doors. In the case of the single sliding panel door, the gasket is mounted on the wall face of the panel near the upper edge and the vertical edges. When the door is in place it is pressed firmly against the door frame to make a positive seal. In some models of sliding doors the face and bottom gaskets are merged to form a single continuous seal. Where the door weight cannot be used to apply pressure to the gasket, as is the case with bi-folding doors, then sweeper gaskets must be employed. Sweeper gaskets are fixed to the undersides of swinging and bi-folding doors and are in constant contact with the floor during door operation. Consequently their

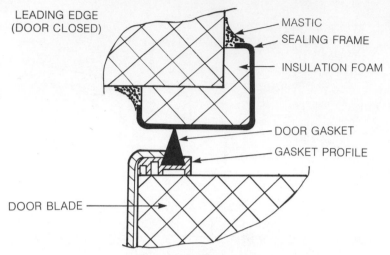

Figure 10.15 Single gasket.

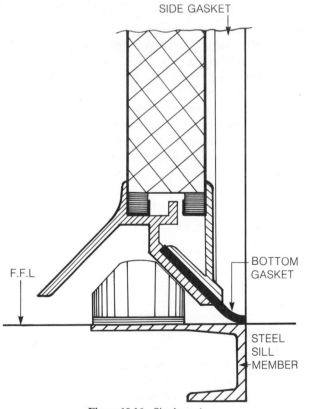

Figure 10.16 Single gasket.

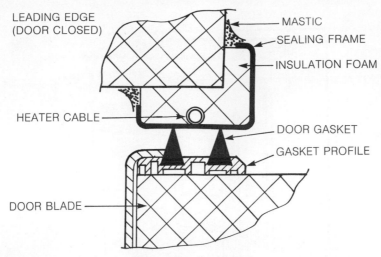

LEADING EDGE
(DOOR CLOSED)

MASTIC

SEALING FRAME

INSULATION FOAM

HEATER CABLE

DOOR GASKET

GASKET PROFILE

DOOR BLADE

Figure 10.17 Double gasket.

rates of wear are greater than those of face gaskets and bottom gaskets, which are usually lifted clear of contacting surfaces during door operation. In the case of step-in freezers these face gaskets can be used to seal all four edges of the door panel, but if the door sill is at floor level a sweeper gasket must be employed on the underside. Bi-folding doors need special gaskets to seal the joints between adjoining panels.

Attention should be paid to methods of fixing gaskets, enabling them to be easily replaced in the event of damage. The gaskets of sliding doors, which drop into position in their final closing movements, will not wear as quickly as those on, say, sliding doors with wedge action closing, or sweeper gaskets which constantly rub against the floor. This should be taken into account when selecting gaskets.

10.4 Loading dock systems

The majority of traffic to and from the larger cold stores is often long-distance refrigerated transport. During loading or unloading at a conventional open dock, there is complete air infiltration into these wagons. There is also considerable infiltration into the cold store due to the traffic through it, to and from the transport. Infiltration to both the cold store and the transport can be greatly reduced by a loading dock system (Figure 10.18).

To employ such systems the client needs to standardize his transport fleet as far as possible, although the systems permit a slight variation in dock height and load box section. The standard store aperture is 2700 mm wide × 3000 mm high, set about 1500 mm above ground level. An insulated shelter is built in front of the aperture. The floor of the shelter includes a hydraulic ramp

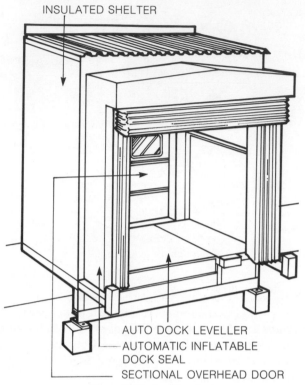

INSULATED SHELTER

AUTO DOCK LEVELLER
AUTOMATIC INFLATABLE
DOCK SEAL
SECTIONAL OVERHEAD DOOR

Figure 10.18   Loading dock.

which will adjust to the height of the vehicle load platform as it varies during loading and unloading. The store aperture is usually closed by a sectional overhead door, mounted internally. When a vehicle backs up to the dock, a flexible cushion inflates and seals to the sides and top of its body. The front of the dock carries two rubber bumpers to protect it from impact with heavy vehicles. When the vehicle docks the store door is opened and the cushion inflates, so that the vehicle and the store are sealed from air infiltration. Loading and unloading can then proceed without the inconvenience of opening and shutting doors and without the huge losses incurred by infiltration. It is not possible to seal the small space between the hydraulic ramp and the tailboard of the vehicle, so in fact there will be a little continuous leakage, and some low voltage heating may be required to prevent icing on the ramp.

Another piece of useful equipment where vehicles cannot back up to a dock of convenient height is the hydraulic lift table (Figure 10.19). This table can be used as a bridge between a dock and a vehicle. It can be sloped if necessary, or it can be used as a lift to raise heavily loaded trollies from floor level to vehicle level, thus cutting down goods handling time.

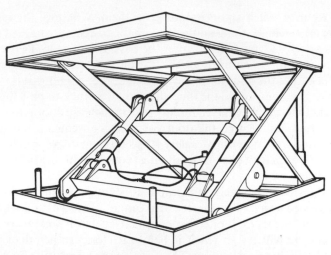

Figure 10.19 Hydraulic lift table.

10.5 External door controls

There is a variety of methods by which cold store doors can be opened without requiring the truck driver to dismount. Two of the most common are the pull-cord actuator and the radio-controlled actuator. For radio control, each truck is equipped with a transmitter and each door is fitted with a receiver which actuates the door opening mechanism. Radio controls do not oblige the truck driver to stop before opening the door. There is a temptation for drivers to leave actuation until the very last instant while they drive at full speed towards the door—judging the door opening as finely as possible can come to be an amusing game in a boring day's routine—with consequent risks of collision damage to the door. Problems arise also if the store has multiple doors. It becomes expensive to supply each truck with sufficient transmitter channels to activate all doors independently. There is also the possibility of a driver selecting the wrong channel whilst careering towards a door.

The virtue of the pull-cord actuator is supposed to be that it obliges the driver to stop in order to pull it. However, drivers rapidly acquire an ability to pull actuator cords without stopping. This has already been discussed in Section 10.3 where it was argued that the best location for an actuator was that which sited the truck as close as possible to the door. The company installing the door should be advised before positioning a pull-cord actuator, otherwise, as often happens, the actuator is simply fastened to the most convenient steel girder.

Once the door has been opened it needs to be shut as soon as possible. Not all truck drivers can be relied upon to do this, especially if it means stopping the truck to do so. Experience shows that drivers will often leave doors open whilst they are inside a cold store, perhaps through laziness or perhaps for the feeling of security which an open door gives them. Whatever the reason, the infiltration losses can be tremendous. Automatic systems are needed to ensure that doors are closed immediately after the passage of traffic. Two automatic systems are available to ensure door closure: the timer and the proximity loop.

The timer causes the door to close a fixed time after opening. It is usually set to err on the generous side in the time allowed for a truck to pass through the door: should the truck be delayed for any reason then the door is likely to shut before the truck has cleared the doorway, with consequent damage to the door. Most doors carry sensitive safety strips on their edges which will cause them to reverse rapidly if there is the slightest contact with the sides of the load or the truck.

Safety edges are extremely vulnerable to damage from trucks and are frequently ripped away. Also, a safety strip cannot protect a door which closes in front of a moving truck. It often happens that if a driver sees an open door in front of him, he will try to reach it before it closes rather than stop and reactivate the opening mechanism. There is also the temptation for a driver to follow another truck through an open door, and if the door is controlled by a timer the probable result will be a costly collision.

Proximity loops are used in conjunction with pull-cord actuators. Pull-cord actuators are used to initiate door opening and proximity loops are used to initiate door closing. The loops are embedded in the floor on either side of the cold store door. Once the pull-cord actuator has initiated door opening the door will not close until the truck has cleared the proximity loop on the opposite side. This ensures that the door will not close on a stationary truck or on a following truck. The system is especially useful where tractor train transport is used.

Other detectors which have been tried are photoelectric and ultrasonic devices. Photoelectric devices are not completely reliable since in certain climatic conditions, fog will form in the vicinity of cold store doors and interfere with their operation. Ultrasonic devices require considerable maintenance.

A complete opening and closing system can be provided by proximity loops with two loops: one outside the door and one inside the door. As a truck approaches the door it crosses the first loop which will open the door, and after the truck passes to the other side it crosses the other loop which will close the door.

The ordinary proximity loop cannot differentiate between a truck which is approaching a door and one which is merely passing in front of it. This can be a wasteful nuisance in a multi-door loading dock. Direction-sensitive loops can be devised which differentiate between a truck approaching a door and one merely crossing in front of it; such loops tend to be expensive but may well be worth the extra cost.

Where pedestrians use cold store doors frequently, it pays to fit the doors with a partial opening control. These controls operate with push buttons which open the doors only sufficiently for pedestrians to walk through comfortably and then closes them afterwards. It may be preferable to use a simple lever to close the doors, although one must trust the pedestrian to use these.

Some stores are entered through vestibules which are intended to prevent excessive infiltration. It would appear logical to provide a vestibule with two interlocked doors, one at each end, to always keep a closed door between the cold store and the outside. But this obliges the driver to stop his truck and wait for the first door to close behind him before he can open the second. Such small delays prove irritating, and tempt drivers either to remove the interlock or to wedge the inner door open. Cases of such interference have been many. Experience shows that it is better to have no interlock and accept the small amount of infiltration which occurs in the few seconds when the two doors are open together.

10.6 Air curtains

Air curtains are two-dimensional jets of air projected across doorways in either the downward or the transverse direction. Their purpose is to prevent the ingress of ambient air and the egress of cold air. Research has shown that they can perform this task with efficiencies in the order of 85% to 90% [12, 16, 26, 29, 32] without obstructing visibility or traffic flow. Efficiency is defined as

$$\eta_{AC} = 1 - \frac{\overset{\circ}{V}_{AC}}{\overset{\circ}{V}}$$

where $\overset{\circ}{V}_{AC}$ is the rate of infiltration with the air curtain operating and $\overset{\circ}{V}$ is the rate of infiltration without the air curtain (calculated from Equation 10.9).

With the efficiency levels indicated, infiltration through the air curtain must be 10% to 15% of that through a continuously open and uncurtained door. This is unacceptable outside the working day or during slack periods when traffic density is low. But when the traffic density is high the air curtain infiltration rate can be less than that through frequently opened doors. A rule of thumb guide suggests that the air curtain becomes economical when total door-open time is greater than 1 hour in 24. Air curtains are always used in combination with doors so that the doors can be shut firmly and the air curtain fans switched off during non-working hours or slack periods.

Two designs of air curtain are available—recirculatory and non-recirculatory, of which the simplest and cheapest is the non-recirculatory type. Several writers have questioned whether the much greater cost of the recirculatory type is sufficiently justified by their superior performance and certainly they are never used in cold stores.

The main part of the air curtain apparatus is a centrifugal fan with forward swept blades, driven by a single-phase or three-phase electric motor. In the non-recirculatory type, the fan and motor are mounted axially in a casing or duct, and the assembly is fitted above the front of the cold store doorway. Ambient air is drawn into the fan through an axial duct and expelled radially. The delivery is then projected downwards and slightly outwards at moderate velocity through an adjustable nozzle which runs the entire length of the doorway. The two-dimensional jet produced deflects the warm air stream which tends to penetrate the store at the top of the doorway. The reaction between the two streams also deflects the air curtain vertically downwards or angles it slightly inwards to oppose the flow of cold air which tends to escape at the bottom of the doorway. On striking the floor, part of the jet tends to deflect towards the store and this too opposes the escape of cold air. At the top of the doorway the jet entrains warm air on the outside and cold air on the store side; this attracts cold air towards the top of the doorway and minimizes temperature rises in the vicinity of the door.

The jet nozzle should be set as close to the top of the doorway as possible. Jet angles and air velocities must be carefully adjusted if the air curtain is to achieve high efficiency. If air velocities are too small, the curtain will fail to reach the floor and warm air will penetrate the store and cold air will escape. If the jet velocity is too great, then the air curtain will split on hitting the floor and part of the warm jet will penetrate the store along the floor [29].

The effectiveness of an air curtain in protecting a cold store depends on the rate of supply of momentum: it is proportional to the product of the thickness of the curtain and the square of the air velocity. In turn, the required air curtain stiffness depends on the square of the doorway height. This means that

$$\text{Power to drive fan} \propto \frac{b \cdot y}{d}$$

where d is the thickness of the jet leaving the nozzle.

Door openings should be protected against strong winds which can penetrate air curtains. If air curtains are used then there should be no other unprotected openings in the store. Where a store has a number of air curtained doorways, these should all be set in the same wall.

Air curtains were in vogue in the later 1960s and the early 1970s but are now declining in use in cold stores.

10.7 Internal cold store doors

Cold store complexes often include adjoining areas which are maintained at the same temperature or at different temperatures with high density traffic passing between them. Internal doors or other partitions are often needed to separate these areas. Several types are available: sliding doors, plastic swing doors, panel type swing doors, strip curtains, and roll-up doors. Internal doors are frequently needed in fire walls, compartment walls or to separate main store areas from blast freezers. It may be possible to use air curtains for the same purposes. Where temperature differences between areas are great, the obvious choice is one of the range of sliding panel doors described in Section 10.3. Where temperature differences are small, other, lighter barriers are cheaper and easier to install and maintain. Truck drivers and pedestrians generally prefer clear-view doors—constructed of transparent materials like PVC or polycarbonate—for their obvious safety advantages; it can be unnerving to push through an opaque barrier, not knowing that may be approaching in the opposite direction.

Light sliding doors and bi-folding doors constructed of transparent material such as PVC are available. These carry no gaskets and do not seal hermetically, but they do offer a degree of protection against air infiltration between areas. They are mounted on overhead rails in the same manner as the heavier external doors, but the bearings, rails and door frames are of much lighter construction. It is possible to operate them using radio control, pull-cord actuators or proximity loops, or any combination of the three for greater operational efficiency in high traffic density conditions.

Plastic swing doors made of clear polycarbonate provide tough barriers which can be pushed open easily by trucks or their loads. The lower halves of these frequently are made of neoprene rubber to give added resistance to impact. For heavy duty application, steel buffers can be incorporated to absorb the shock of truck impact. It is recommended that if loads exceed 60% of the door width, then these doors should be fitted with electrical or electro-pneumatic opening devices. The panels are flexible and have to be supported along their tops and sides by L-shaped, light metal frames. Panel frames and door frames are available in stainles steel or heavy duty galvanized steel in either tubular or box section. Some models mount their door panels on panel frame pivots set at the top and the bottom of the panel frame uprights. The heavier duty doors are mounted on specially designed, spring-loaded steel hinges to protect them from impact damage.

Strip curtains are often used to divide working areas. These are wide strips of flexible and transparent PVC, hung from overhead rails. Various designs of fixings are available to facilitate easy replacement of damaged strips. Strip curtain assemblies can be mounted on overhead rails so that they can be moved to one side in the same way as sliding panel doors. Such installations can be opened and closed automatically using electric or electro-pneumatic motors. In operation, strip curtains form flexible barriers through which trucks and pedestrians can easily pass. They do not usually reach floor level and are only moderately effective as infiltration barriers. Several objections have been levelled at their use as a flexible barrier in truck routes:

- They tend to be soiled with dirt and grease from trucks and are difficult to clean
- Strips are easily caught on trucks and loads and torn
- Strips have been known to dislodge loads from trucks

- Strips are prone to age embrittlement—they break and leave dangerous hard, jagged edges

Swinging panel doors provide a higher degree of insulation between areas of differing temperatures. Single panel doors are mainly used for pedestrians, and double panel doors for trucks. Double panel doors can be fitted with steel bumpers to resist the shock of impact from trucks or they may be opened automatically. The door panels are built with timber sub-frames clad on both sides with stainless steel, aluminium or galvanized steel sheet. Polyurethane foam is injected into the door under pressure to give insulation and increased impact resistance. The undersides of the door panels are sealed by sweeper gaskets. Panels surfaces are finished with a coat of vinyl paint. Door frames are usually of timber surfaced with steel sheet, and often finished with vinyl paint.

High speed roll-up doors are mainly useful for internal divisions, but they have been used as external doors for chill stores with temperatures between 0–5°C. The flexible door panels are woven from polyester fibre and have flexible coatings. At the lower end the panel carries a rigid metal bar and at the upper end it is wound around a long overhead roller. A reversible three-phase electric motor drives the roller through a slipping clutch. The flexible sheet and the metal bar are guided by the channelled door frame uprights as the door is raised and lowered. These doors are available up to any size conceivable for cold stores. Operating speeds range up to 1.3 m/s giving a minimum opening time of 5 to 6 seconds. The disadvantages of vertical-opening doors have already been explained, and these are shared by the vertical-opening roll-up doors. Double bi-parting roller doors which open in the horizontal direction are also available. These doors can reach very high opening speeds: one model, of aperture 2 × 3 m, can open in two seconds.

Roller doors can be controlled automatically by all the means described earlier: pull-cord actuators, radio control, and proximity loops; radar, ultrasonic and photoelectric techniques are also used.

10.8 Safety requirements

Very little has been written which gives a guide to the safety aspects of cold store doors.

In the UK, the *Health and Safety at Work Act* lays down various requirements in a general way but gives no exact specifications of the means of achieving safety :

- Page 17, section 29 (1) requires '... safe means of access to every place at which any person has at any time to work'
- Page 27, section 40 (1) requires '... such means of escape in case of fire for persons employed in the factory as may reasonably be required in the circumstances of the case.' The section then goes on to designate the local

fire authority as the examiner of such installations, with the authority to grant a certificate of acceptance.

- Page 28, section 40 (1) states 'All means of escape... shall be properly maintained and kept free from obstruction.'

It seems reasonable to infer that large cold stores should be provided with means of escape apart from the main doors. Such exits could be needed in the event of fire in the vicinity of the main doors. It goes without saying that it should be possible to open such escape hatches or doors easily from inside. It should always be possible to open main doors from the inside by mechanical means in case of power failure. Main doors have been constructed with small pass doors or escape panels built into them. The *Health and Safety at Work Act* insists that all access should be kept clear of obstruction, for obvious safety reasons. For the same reason, stores should always be provided with emergency lighting—especially in the vicinity of doors and escape hatches—so that personnel can locate them in the case of power failure.

EEC legislation, as reported in the *Official Journal of the European Communities* (reference L59 of 5th March 1983) has little to say about cold store doors; its main concern is with hygiene, In *Annex 1, page L59/19 Section 1.c* it states that in rooms where fresh meat is prepared or stored 'doors should be surfaced with hard-wearing, non-corrodible material, and if doors are of wood, with a smooth impermeable covering on both sides'. BS 4434: Part1:1980 Specification for Refrigeration Safety is more concerned with the safety of refrigeration systems, but no mention is made of doors. The *FAO Agricultural Services Bulletin 19/2, paragraph 34* is again concerned with hygiene, specifying hard washable surfaces and the non-use of wood.

The US or Canadian Underwriters Tests deal with the subject of fire resistance in doors. A door to be tested is subjected to naked gas flames on one side for 3 hours. The hot side should rise to temperatures no higher than 1060°C; after 30 minutes, the cold side of the door should not exceed 120°C. At the end of the test, gas flames are removed and the door is sprayed with cold water to test its resistance to thermal shock.

All electrically operated doors should be provided with isolating switches in case of accident. These switches should be mounted on the cold store wall, as close as possible to the doors. It is a good idea to paint them in black and orange stripes to make them highly visible.

10.9   Conclusion and future developments

On reviewing the technology of cold store doors, it can be said that the construction of doors is very sound technically and a very wide choice is available to suit all requirements of the cold store industry.

One major problem is encountered with the types of insulation which are used at present, not only in doors but also in cold store wall panels.

Polyurethane foam and expanded polystyrene have commonly been used because of their excellent insulating properties, low cost and ease of handling. However, these materials represent a very grave risk in the event of fire, in that they will not only burn but may also give off large volumes of lethal cyanide gas. In the future, it is likely that there will be more stringent legislation which will require the use of safe, flame-resistant insulants in cold store construction.

Perhaps the greatest hazard to cold store doors is the FLT. Several suggestions have been made in this Chapter of ways to maintain safe FLT driving and reduce the probability of accidents. In the future, electronic technology may well help. It is perfectly possible to construct proximity loops which, in conjunction with microprocessors, will measure the speed, position and direction of motion of trucks. Such intelligent systems could be used to detect trucks which are heading for collisions. The addition of radio control would enable the detector to apply the brakes of trucks in hazardous situations and bring them to an immediate halt.

In the event of door damage it should be possible to effect repairs quickly and cheaply. Apart from the cost of labour and materials, enormous financial losses can accrue when access is blocked for long periods. One answer to this problem is the use of 'throw-away panels' which can be quickly removed from the panel frames and easily replaced.

References

[1] Giant doors insulate large model cold room, *Industrial Research and Development*, 19th July 1981 **23** (7), 130–133.
[2] High speed roll-up doors cut cooling time and energy costs, *Food Engineering* (1985) **57**, 109.
[3] Special dock doors keep energy costs low, *Food Engineering*, (May 1985) **55**, 175.
[4] Balbach G.C., Hydraulics: a contemporary solution to cold storage door problems, *Dairy Field* (1982) **164** (1), 65–67.
[5] Clement P., Les portes isothermiques pour chambres frigorifiques à basse température, *IIR Bulletin* (1960) **3**, 509–512.
[6] Coleman R.V., Doors for high rise refrigerated storage, *ASHRAE Transactions* (1983) **89** (113), 762–765
[7] Fleming A.K., Design aspects of cold stores, *Transactions of the Inst. of Professional Engineers, New Zealand* (1985) **12** (3), 139–146.
[8] Gosney W.B., Heat and enthalpy gains through cold room doorways, *Inst. of Refrig.*, Paper presented 4th Dec. 1975.
[9] Harbord F., Mechanically operated, sliding, insulated door with positive seal, *IIR Bulletin* (1960) **3**, 479–483.
[10] *Guide to Refrigerated Storage IIR* (1976).
[11] Knapp W., Doors in Cold Stores, *Tiefkuhlpraxis Int.* (Jan. 1971) **12** (1), 25–27.
[12] Michael W.R., Air curtains for use in cold stores, *IIR Bulletin* (1960) **3**, 489–495.
[13] Pau J., Insulated doors for large cold rooms, *Ingenieur Constructeur* (April 1979) **68**, 22–25.
[14] Pham Q.T. and Oliver W.D., Tracer techniques and data interpretation for air infiltration measurements in cold stores, *Proc. 16th Int. Congr. of Refrig., Paris* (1983) **4**, 61–66.
[15] Pham Q.T. and Oliver W.D., Infiltration of air into cold stores, *Proc. 16th Int. Congr. of Refrig., Paris* (1983) **4**, 67–72.
[16] Laudermire K.A., Cold store door seals, *Frozen Foods*, Dec. 1982.
[17] Szeluto J. and Rojewski S., Cold store doors, Chlodictwo, Jan. 1979.

[18] Wilder C.M.G., Doors for refrigerated warehouse, ASHRAE Seminar Boston (USA), 1975.

[19] Atkins W.S. and Partners, Epsom, Study of refrigerator door losses, Study carried out on behalf of Clark Doors, July 1982.

[20] Tojanowski T.J. and Rubnikowicz A., Cutting the cold loss through cold store openings with vertical air curtains, *Luft und Kaeltechnik* (1981) **17** (3).

[21] Ullner F., New generation of cold store sliding doors in the GDR, *Luft und Kaeltechnik* (1987) **23** (1), 42–45.

[22] Van der Hoek W.T., Losses due to opening of doors in cold stores, *Koeltechnik* (*Klimaat*), (June 1981) **74** (6), 125–127.

[23] Van der Spek W., Sliding door systems for cold chambers, *Refrigeration and Air Conditioning* (June 1973).

[24] Markus J.J.G., Sliding Doors for Cold Rooms, *Koeltechnik* (*Klimaat*) (Feb. 1969), (2), 41–44.

[25] Wright C., Selecting energy efficient storage doors, *Food Engineering* (1984) **56** (2), 148–150.

[26] Hayes F.C. and Stoecker W.F., Design data for air curtains, *ASHRAE Transactions* (1969) **75** (2), 168–180.

[27] Niculita P., Purice N. and Coibanu A., Experimental studies on air curtains in single storey cold stores, *Lucr. Cercgtinst. Chim. Aliment* (1984) **14**, 67–76.

[28] Longdill G.R. and Wyborn LG., Performance of air curtains in single storey cold stores, *Paper 15th Intr. Congr. of Refrig.*, Vienna (1979), 77–86.

[29] Asker G.C.F., What, where, and how of air curtain systems, *Heating, Piping and Air Conditioning* (June 1970) **42** (6), 56–62.

[30] Georges A., Heated door frames for cold rooms, report, *IIR* (*Paris*), (1982) **2**, 89–93.

[31] Van Male J., A new vertical air curtain design for cold stores, *Proc. 16th Int. Congr. of Refrig., Paris* (1983) **4**, 73–82.

[32] Tamm W., Airflow within air curtains to protect cold rooms, *Proc. 11th Int. Cong. Refrig.*, Munich, (1963) **1**, 1025–1033.

[33] Van Male J., Optimum design for air curtains for cold rooms, *Proc. 15th Int. Cong. Refrig.* (Venice 1979) **4**, 89–93.

11 Fully automated cold stores

R. TAYLOR

11.1 Introduction

There has been a dramatic increase in the number of frozen food items which are taken home and stored in the refrigerator. During the early 1960s the frozen product section of a supermarket was typically one row—now it can be as much as one third of the area of the store. Frozen food manufacturers are competing to produce more and more exotic products for the microwave oven. In the USA, over 50% of all households use a microwave, and more than 400 new products arrived in the supermarkets in the first half of 1987.

With these trends, it has been difficult for manufacturers and food distribution centres to keep up with the demand for refrigerated space. The time has come to make major improvements in productivity, and to reduce manpower through automation to get the products to the consumer at the lowest possible price without sacrificing quality.

The freezer warehouse, however, is one of the most difficult environments for man and machine. Man-hour productivity levels are greatly reduced at $-10°F(-23.3°C)$ and below, and equipment costs are higher due to the high reliability factors and preventive maintenance programmes required.

Warehousing is an unavoidable cost of doing business, as is the cost of transportation of products to the seller. Neither of these costs has anything to do with the manufacturing cost of the products to be sold, but they play a very important part in the final cost to the end user. Most companies spend very great amounts on developing new techniques to reduce the cost of the manufacturing process of a given product to stay competitive in the market, but are slow to realize the savings that can be gained from automating their warehousing and distribution. If you need to reduce the cost of a product by 10%, it is easier to find the savings in the material handling area—which may be 65% of the overall cost—than in the manufacturing portion.

The basic functions of the distribution centre have not changed over the last forty years. It must still:

- Receive goods (palletized/unpalletized)
- Store goods
- Ship goods (by pallet, by case, by customer's orders)

Years ago, industries had relatively few manufacturing facilities and in order to get the maximum efficiencies out of manufacturing it was necessary to mass-produce large lot volumes. Hence, large warehouses were required to store products for long periods of time and then transport them across the whole country to their final destination or to a sub-distribution centre.

In the early post-World War II era, the only way to move large volumes of material was by FLTs. These worked generally in large 10–12 ft (3–3.7 m) aisles and stored pallet loads of product at heights of up to 12–20 ft (3.7–6.1 m). These heights were basically determined by general building construction limits and the design limitations of gasoline-powered FLTs.

The freezer storage section of this sort of warehouse was just a small corner of the building, similar to a large meat locker. A single door would be the only access to it. The height of the freezer would be only 12 ft (3.7 m). The area would be serviced with conventional pallet drive-in rack and lift trucks. There was no humidity control, and ice would form on all surfaces, including the freezer door, preventing a good seal, and leading to an increase in temperature.

Order picking was done either in the freezer with small hand carts, or a pallet was brought to the shipping dock, cases picked and the remainder of the pallet returned to the freezer at a time convenient for the FLT operator. This meant that the quality of the frozen products varied greatly, sometimes to the point where the product was completely spoiled and had to be thrown away. In the United States, this became a problem to the Food and Drug Administration (FDA), and stringent rules were enforced to prevent sub-standard frozen products arriving on the supermarket shelves.

11.2 Improved technology and the development of automated storage

Over the years, the technology of FLTs has improved greatly to the point where most warehouse trucks are now battery-powered. The large aisles have been reduced in size and the height restrictions have improved to the point where some man-aboard trucks are now surpassing 50 ft (15 m) storage levels.

There are many different types of man-aboard trucks:

- Counterbalance   —aisle 12 ft (3.66 m)
- Reach   —aisle 9 ft (2.74 m)
- Straddle   —aisle 9 ft (2.74 m)
- Side reach   —aisle 8 ft (2.44 m)
- Turret   —aisle 6 ft (1.83 m)
- Double reach   —aisle 6 ft (1.83 m)

Although FLTs continue to be used throughout the industry, they have some basic inherent problems, due to the fact that they are operator-driven vehicles:

- Design is not specifically for freezers.

- Relatively short life span (5–7 years on average)
- High maintenance costs (hydraulic seals, etc.)
- Reduced battery life
- High manpower requirements
- The operator has no protection against the environment

Also, they may cause

- Produce damage
- Rack damage
- Equipment damage
- Reduced productivity

The advantages of the man-aboard FLT warehouse in distribution centres are:

- Low initial cost
- Expandable
- Easy to use with various types of rack
  —Pallet rack
  —Drive in
  —Gravity flow
  —Power rack
  —Shelving
  —Block storage

The disadvantages are:

- High labour content
- Slow and inefficient
- Limited inventory control accuracy

During the mid 1950s in Europe, when companies did not want to change their location due to the high cost of land and needed more manufacturing space, the use of high-rise storage systems served by stacker cranes became wide-spread.

The first systems were manually operated at the 30–50 ft (9–15 m) height ranges. Many were ceiling-mounted, but floor-running and top-guided systems gradually became more common, and indeed 99% of stacker cranes are of these types today.

During the latter part of the 1950s these stacker cranes were semi-automated; the operator was taken off the crane and positioned at the end of the crane aisles. He would operate the cranes by pushing a series of buttons on a control panel on the back of the crane and the crane would do a 'store and retrieve' cycle. It was also possible for the operator to operate two or three cranes at the same time. It did not take long for the push buttons to be replaced by card readers where a standard 80-column card would be punched with a product and storage location and the cranes would then perform their functions from the instructions on the punched card. The next logical step was to take the card reader off the cranes and install a console remote from the cranes, and, from a single point, one operator controlled the material movement.

This was the real start of the *automated storage system* (AS/RS). It quickly found its way to North America and has since been used in every conceivable application, from plastic washers to steel coils, at heights up to 110 ft (33.5 m) and temperatures from −40°F to 100°F(−40°C to 38°C).

Construction of frozen food distribution warehouses mushroomed in the mid 1960s; more sophisticated techniques of materials handling were also developing. Automation and computerization, the fruits of modern technology, were applied to speed the flow of merchandise from manufacturing to warehousing to consumers. Speed was the key in unloading products, in storing items, in order selection, picking, and loading delivery trucks with minimal change in product temperature.

Formerly, storage space had been the fundamental concern in the minds of warehouse planners and builders. Now, efficiency of the materials handling system became the focal point. Warehouses were now being designed with function in mind first, placing a building or skin over the material handling solution. The length of picking lanes, aisle widths, load heights—in short, the physical dimensions of the working area in the distribution centres—had to be planned to handle the anticipated requirements of product flow with maximum ease. In other words, warehouses were designed according to their desired productivity and the type of materials handling system that would be best suited to achieve it. Finally, these facilities had to be constructed with the utmost flexibility to allow for expansion in keeping with the needs of a rapidly growing market. In the private warehouse sector, it was automation—or more often, mechanization coupled with semi-automation—to which manufacturers, wholesale distributors, and retailers looked for speedier and more economical methods of operation in the 1960s.

Among frozen food processors, the Kitchens of Sara Lee production and distribution complex in Deerfield, Illinois, built in 1964 at a cost of some $30 million, still stands as a landmark in automated warehouse construction. Sara Lee was the first frozen food packer to use the 'high-rise' concept for storage. Measuring 300 ft (91 m) long and 180 ft (55 m) wide, the freezer stored 45 ft (14 m) high. Five stacker cranes serviced 5310 pallet openings at −10°F (−23.3°C). Under control of the computer and a separate emergency back-up control centre, cases of frozen baked goods flow in from the bakery on conveyor lines, at rates of up to 60 per minute. In a pallet loading area, the cases are arranged in interlocking unit load layers. The layers are picked up by overhead vacuum lifters and put on pallets. Completed loads flow past a photocell check point into the −10°F(−23.3°C) warehouse, a freezer bigger than a football field. At the computer's instruction, the loads will be shunted onto a two-pallet accumulation conveyor sections at any one of ten aisles, to be picked up by a computer-directed stacker crane and placed in racks. The computer keeps track of loads by product, location, and time in storage, and maintains a balanced load in the aisles on a random basis. Since the computer knows where every item is, there is no need for specific locations by product. FIFO picking is controlled by the computer's memory. The computer tells the proper stacker crane to take out whichever pallet load is due to be shipped. The stacker then automatically takes the load to the main conveyor which then takes it to the truck or rail docks.

The high-rise approach really expanded after 1968 when Stouffer Foods constructed the freezer portion of its Solon, Ohio, plant to a height of 120 ft (37 m). This was a daring innovation. In its multilevel design and its high degree of automation, Souffer's new freezer storage facility represented the most radical departure from the prevailing concept of single-level construction for refrigerated warehouses.

The principle of the high-rise freezer, often coupled with fixed-rail stacker cranes, was soon adopted by a number of well-known distributors, manufacturers, food chains, and public warehouse firms.

A number of other companies, although they did not go as far as stacker crane installations, did rely on advanced concepts of mechanization and semi-automation in high-rise or semi-high-rise low-temperature warehouses covering the gamut of processing, storage and distribution. One example was Lucky Stores, Inc., of Dublin, California, which in 1977 opened an automated selection gravity feed/conveyor belt installation at its southern division distribution centre in Buena Park, California.

In at least one instance in recent years, warehouse improvements have come about in an unusual way. Following a fire at its Harrington, Delaware freezer in 1979, Burris foods made significant changes to the method of operation of the automated facility—for example, by replacing four single-pallet hydraulic stacker cranes with four electromechanical two-pallet stacker cranes—which almost doubled the facility's throughput.

Another company which doubled frozen food warehouse output through automation was H.E. Butt Grocery Company. New technologies in order selection also enabled the chain to halve the manpower necessary for its San Antonio, Texas, freezer.

In 1982, Giant Food Inc., a Landover, Maryland chain, opened a 130 000 ft^2 (12 077 m^2) frozen food distribution area featuring an SI Ordermatic system capable of picking 3600 cases per hour. Rotelle, Inc., a Philadelphia area wholesaler, which had moved its food service operation into a new complex in West Point, Pennsylvania the year before, shifted its retail distribution to the new site where four fully automated Interlake cranes handled the merchandise.

11.3 Additional influences

Meanwhile, all segments of the low-temperature warehousing industry began to tackle the increasingly critical problem of energy. In the face of soaring power costs and fuel shortages, the International Association of Refrigerated Warehouses (IARW) in 1974 published a list of 39 steps designed to conserve warehouse energy without jeopardizing frozen food quality. The industry group continued to pursue energy-efficient construction and operational guidelines, aided by its affiliate The Refrigerated Research Foundation, and at its 1976 convention reviewed ways of improving the energy efficiency of older facilities. Later that year, IARW issued a loose-leaf energy conservation manual to its members, covering operations, management, energy audits and other related topics. Other industry groups followed suit, and the effects were widespread. By 1977, 37.4% of warehouse operators—public and private— reported that they had implemented specific design and/or operational changes based on energy considerations.

Another major concern was increased government scrutiny of the food industry. Warehouse operators were quick to sharpen their housekeeping skills and practices as the FDA stepped up its warehouse inspection programme. Warehousing management had been especially alert to this subject since a 1975 US Supreme Court ruling that top corporate executives could be held personally liable for FDA violations in warehouse sanitation cases.

The National Frozen Food Association was especially active in this area, having revamped its *Certified Full-Service Wholesaler Programme* in 1972 to incorporate sanitation inspection by the American Sanitation Institute. Then in 1982 NFFA upgraded the inspection programme for distributor and warehouse members and renamed it *Certificate of Excellence: Service, Sanitation and Safety.*

Freezer warehouse growth was strong during the latter half of the 1970s, as the frozen food industry anticipated significant acceleration in its sales after the depressed levels suffered during the 1974–75 recession, an expectation that was to prove justified when the frozen food industry rebounded with a robust performance in 1977.

Before the high interest rates which were introduced in 1980, the demand for warehouse space seemed to override concern about high construction costs, the cost of money and uncertainty about the frozen food sales outlook. A 'mini-boom' in warehouse construction was attributed, in part, to the revival of projects that had been deferred for several years by cost and other concerns. By 1979, an inflationary policy seemed to be operating: better to build the facility today because it will cost more tomorrow. Some were financing the construction internally, others had planned and financed projects when money was cheaper. Thus, low-temperature construction in 1979 continued apace, underscoring the basic reality that if the new facility was essential to stay competitive, the importance of interest rates became secondary.

During a 1981 workshop, Richard Wagner, owner/founder of World Unlimited, predicted that warehouse construction would increase in a year or two because 20–25% of cold storage space was obsolete. He cited energy inefficiencies, improper handling capabilities, and unsanitary conditions as underlying the obsolescence. The period between 1983 and 1986 certainly confirmed his predictions as the number of high-rise warehouse freezer facilities more than doubled.

11.4 Automated storage/retrieval systems: some first considerations

Objectives

The purchase of any type of equipment must first be justified. In the case of the automated warehouse it is generally justified by comparing a manually controlled FLT system with the planned AS/RS. Unfortunately there are many intangibles connected with automation that are difficult to put a value on, e.g. absolute inventory control.

Before trying to justify an automated storage/retrieval system (AS/RS) the requirements of an automated storage system should be considered. Among others these include:

- Improved material flow between manufacturing and warehouse
- Relocation to give more manufacturing space
- Expansion of warehouse size (more products, more volume)
- Combined warehousing functions
- Minimal equipment and product damage
- Reduced handling procedures
- Storage cube standardization
- Improved stock rotation
- Control and reduce inventories
- Reduced operating costs
- Improved operational environment
- Reduced pilferage
- Increased productivity
- Compliance with OSHA (Occupational Safety and Health Act) and other governmental agencies
- Utility reductions
- Controlled product life cycles
- FIFO product
- Improved customer service

Load

The next step is to define the product to be handled. This is the *most important* step because it affects all other components of the selection of material handling equipment.

Things that must be known about the load to be stored and conveyed are:

- Pallet size–how many?

- Pallet construction—will it convey?
- Pallet deflection—at centre point (maximum load)
- Load size—width, depth, and heights (are these variable heights, if so what are the percentages of each?)
- How is it stacked on the pallet?
- Will it expand if stored for long periods of time?
- Is load strapped, shrunk-wrapped, etc.?
- Is the load on a slip sheet?
- Load maximum, minimum, and average weights
- Does the product have a life expectancy?
- How is the load identified (label, barcode, etc.)?
- How many SKUs (stock keeping units)?
- How many pallets of each SKU?
- Is the product to be case picked or handled as a unit load (pallet load)?
- How many cases per pallet or SKU?
- Is a slave pallet (generally a plywood slab captive to the storage system to create uniformity) required?

Activity

After collecting all the information about the load to be handled, it is necessary to create a flow diagram of where the loads come from, and where they are going to. Generally a unit load AS/RS performs better and faster when storage and retrieval functions are from a common end of the crane aisle, and systems are generally laid out in this way. But sometimes the physical layout of an existing facility does not allow this, and so both ends of the rack structure are used for input or output, or sometimes for both functions. The use of conveying methods other than FLTs can transport products to adjacent order-picking areas and shipping docks in order to try and use a common aisle end.

The flow chart, along with activity rates of the different products to be stored and retrieved, should be broken down into an hourly rate so that the correct type of crane, conveying system, and respective speeds can be established.

In many cases production is a three-shift operation whereas shipping usually happens through an 8–10 hour period of daylight, therefore a balanced distribution system generally has very high activities during the shipping cycle, while the other two shifts of production are lower. The ideal world would have balanced shipping and receiving, but this is seldom the case. It is expensive to design a system for a peak activity. If it is possible to spread a peak period over an extra couple of hours, that could help in reducing the cost of the equipment and assist in the justification of the project. Automated equipment can run without an operator being present, therefore material can be stored or retrieved during 'off' hours without running up labour costs.

Different types of AS/RS

Figure 11.1 shows a spectrum of the different types of AS/RS vehicles that are available with today's technology and have had many years of successful operating experience. The graph relates storage density with system throughput in very general terms by placing vehicle types closest to the area for which they are best suited. This is simply a guide to help the user get a feeling for the types of systems available. A comparison of some examples may also be helpful. Consider first a system with multi-aisle, one machine, pallet entry forks, and transfer car, with 20 single transactions per hour. This type of system has a very low throughput but has 100% random storage availability. The storage density is low compared to deep lane storage systems which are called high-density storage but store a small number of SKUs. In contrast, a machine in each aisle, twin shuttles, with 40 dual transactions per hour has a very high activity rate compared to a single shuttle vehicle and can be used as a 100% random and high-density storage system.

As a good rule of thumb, AS/RS systems should have a 6 : 1 ratio of length:height of the rack structure to be most cost effective.

Different types of cranes have varying throughput capabilities. The following crane activity rates can be used as a guide:

- A   Single deep AS/RS   25 dual commands/hour
- B   Double deep AS/RS   22 dual commands/hour
- C   Single deep/twin shuttle AS/RS   40 dual commands/hour
- D   Double deep/double wide aisle   45 dual commands/hour

Obivously A is the only system that works with purely 100% random storage. C could be used for random storage but the activity would be reduced especially if product comes to the pick-up station as mixed SKUs. Machine types B and D are high-activity, high-density storage system vehicles.

If the number of storage aisles is greater than the number of cranes required to service the system, a transfer car can be employed which can be installed at the rear or the front of the system (Figure 11.2). Crane activity rates should be reduced by 10% when using transfer cars.

If a transfer car is installed at the rear of a system, one transfer car can handle several cranes. If they are installed at the front end then conveyorized pick-up and delivery stations can be installed on top of them. Under this scenario, one transfer car is required for each crane. This is a very cost-effective system if the system throughput is low compared to the number of aisles required for storage.

Figures 11.3 and 11.4 and Table 11.1 summarize the recommendations of the Material Handling Institute (AS/RS Section) for the calculation of cycle times, for single or double deep storage systems, with different commands.

Three of the total cycles per hour for an AS/RS machine should include shuffle cycles, in order to indicate the throughput degradation compared with

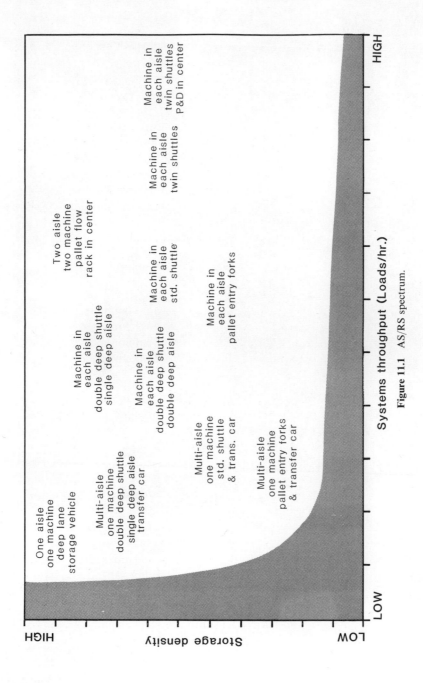

Figure 11.1 AS/RS spectrum.

Figure 11.2 Storage aisle with transfer car.

dual cycles without shuffles. Actual system operation may require a varying number of SKUs in storage, and the system operation algorithms.

In cases where 1/2 or 3/4 of the storage locations (horizontal, vertical) equals a fractional number, round off to the next higher (longer) storage locations.

Table 11.2 gives the cycle times that are possible with a crane work-ing a single aisle having a load size of 48 in deep × 40 in wide × 48 in high (1.2 m × 1 m × 1.2 m) with speeds of hoist 60 ft/min (18.3 m/min), travel 360 ft/min (110 m/min) and forks 60 ft/min (18.3 m/min).

START AT "HOME" WITH PICKUP.
RETURN TO "HOME" EMPTY.

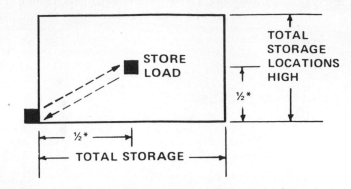

"HOME" AT END OF SYSTEM AND
AT LEVEL OF 1ST STORAGE LOCATION.

Figure 11.3 Computing cycle time for a single command AS/SR.

START AT "HOME" WITH PICKUP.
RETURN TO HOME FOR DEPOSIT.

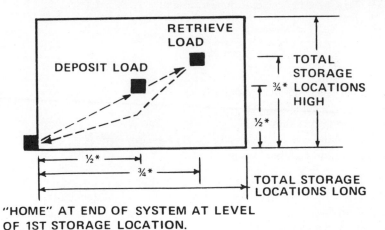

"HOME" AT END OF SYSTEM AT LEVEL
OF 1ST STORAGE LOCATION.

Figure 11.4 Computing cycle time for dual command AS/RS.

11.5 Racks

Over the past 20 years, every conceivable rack configuration has been used in conjunction with freezer AS/RS systems. In order of popularity, they are:

Table 11.1   Methods of computing cycle times.

SINGLE COMMAND, SINGLE CYCLE

Step 1   Retrieve a load at pickup station (single deep stroke cycle)
Step 2   Travel with the load to storage location 1/2 the number of storage addresses along the aisle, and up 1/2 the number of vertical storage locations in the aisle
Step 3   Deposit a load in storage location (average time between single and double stroke cycle)
Step 4   Return to pick-up position
Step 5   Deposit load

DUAL COMMAND, DUAL CYCLE

Step 1   Pick up a load at pickup station (single deep stroke cycle)
Step 2   Travel with the load to storage location 1/2 the number of storage addresses along the aisle, and up to 1/2 the number of vertical storage locations in the aisle
Step 3   Deposit load in storage location (average time between single and double deep stroke cycle)
Step 4   Travel without a load to a storage location 3/4 the number of storage locations in that aisle and up 3/4 the number of vertical storage locations in that aisle
Step 5   Pick up a load in storage location (average time between single and double deep stroke cycle)
Step 6   Return to delivery position with the load
Step 7   Deposit load (single deep stroke cycle)
Step 8   Carriage vertical travel to pickup position (applicable only on multilevel P&D stations)

DUAL COMMAND, SHUFFLE CYCLE

The shuffle cycle (when required) will replace Step 5 of the dual-command cycle.

Step 1   Pick up a load in storage location (single deep stroke cycle)
Step 2   Travel with the load three storage locations (three bays) horizontally and two storage locations (two levels) vertically
Step 3   Deposit load in storage location (double deep stroke cycle)
Step 4   Return travel from the position outlined in Step 2 to position in Step 1
Step 5   Pick up a load in storage location (double deep stroke cycle)

- Single-deep stacker or pallet rack
- Double-deep stacker rack
- Single-deep pallet rack
- Deep lane gravity flow rack
- Deep lane pushback rack

In all rack systems, it is important that there is sufficient clearance in all directions around the load when stored in the rack and when travelling down the aisle on board a crane. Many systems in the past have failed due to trying to squeeze an extra inch out of each opening to make the building shorter and narrower. Freezer systems especially need that extra clearance as frozen products have a tendency to slip and move. Decreased clearance is poor planning as cost savings in building construction do not offset the number of rejected or hung-up loads, which greatly reduce the operating efficiencies

Table 11.2 Theoretical throughputs of a crane working a single aisle: cycle times with single and dual command.

No. Levels	Number of bays (loads)								
	24	30	36	42	48	54	60	66	72
4	54 / 31	51 / 29	48 / 28	45 / 26	42 / 25	40 / 24	38 / 23	36 / 22	35 / 21
5	54 / 31	51 / 29	48 / 28	45 / 26	42 / 25	40 / 24	38 / 23	36 / 22	35 / 21
6	54 / 31	51 / 29	48 / 28	45 / 26	42 / 25	40 / 24	38 / 23	36 / 22	35 / 21
7	51 / 29	51 / 28	48 / 28	45 / 26	42 / 25	40 / 24	38 / 23	36 / 22	35 / 21
8	51 / 29	51 / 28	48 / 28	45 / 26	42 / 25	40 / 24	38 / 23	36 / 22	35 / 21
9	45 / 26	45 / 26	45 / 26	45 / 26	42 / 25	40 / 24	38 / 23	36 / 22	33 / 21
10	45 / 25	45 / 25	45 / 25	45 / 25	42 / 25	40 / 24	38 / 23	36 / 22	35 / 21
11	40 / 24	40 / 24	40 / 24	40 / 23	40 / 23	40 / 23	38 / 23	36 / 22	35 / 21

1 CMD / 2 CMD

originally conceived. To get the maximum reliability out of a crane, the following clearance dimensions (Figure 11.5) are commended:

- Aisle running clearance (load depth + 8 in [+ 203 mm])
- In-rack side clearance single deep (+ 3 in [+ 76 mm] per side)
- In-rack side clearance double deep (+ 4 in [+ 102 mm] per side)
- In-rack side clearance flow rack (+ 4 in [+ 102 mm] per side)
- Rack depth (load depth + 1 in [+ 25 mm])

With the increasingly high cost of refrigeration, users and suppliers are trying to reduce the storage cube for each pallet stored to maintain their competitiveness. The original high-rise freezer systems that were installed used single-deep or double-deep rack structures. In the late 1970s a more

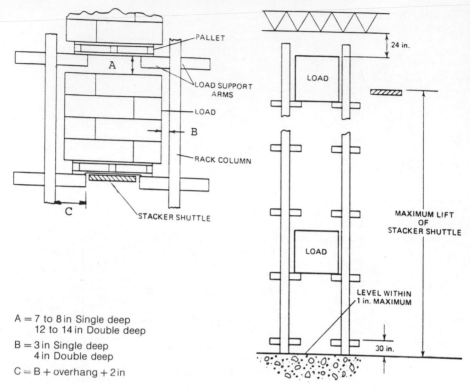

Figure 11.5 Rack clearance.

sophisticated system was utilized using gravity rollers either as a FIFO flowthrough or as a FILO push back, deep lane storage system. This has greatly reduced the storage cube and deserves a little discussion as these systems will become more and more popular in the years to come.

Live pallet storage rack

Live pallet storage offers the following advantages over previous storage systems:

- Each vertical and horizontal storage lane can store a separate SKU.
- Occupancy levels are as high as 85% to 90%
- Minimal, numbers of aisles, aisle width, and travel within the warehouse
- Unlimited storage heights when used with S/R machines
- Product and equipment damage virtually eliminated
- Improved security
- Guaranteed FIFO

The storage systems space allocation chart (Table 11.3) shows the square

Table 11.3 Storage systems space allocation chart, based upon 160 ft × 120 ft area for storage (19 200 ft²), 40 in × 48 in × 48 in high pallet load.

Building height	Material handling equipment	System	Type	Total pallets	Pallets high	Aisle size (ft)	No. of aisles	Sq. ft. per pallet	Store vol. ratio
20 ft	CB L Truck	Selective	P.R.	1920	4	11	8	10	1:3.75
	Narrow aisle L.T.	Selective	P.R	2400	4	$6\frac{1}{2}$	10	8	1:3
	CB L truck	Hi-density	D/I	3360	4	11	4	5.7	1:2.15
	CL L truck	Hi-density	P/flow	3600	4	11	3	5.3	1:2
	CB L truck	Hi-density	Push BK	3600	4	11	3	5.3	1:2
40 ft	Narrow aisle trucks	Selective	P.R.	4800	8	$6\frac{1}{2}$	10	4	1:3
		Hi-density	P/flow	8160	8	$6\frac{1}{2}$	3	2.3	1:1.76
		Hi-density	Push BK	8160	8	$6\frac{1}{2}$	3	2.3	1:1.76
60 ft	AS/RS	Selective	P.R.	8640	12	$4\frac{1}{2}$	12	2.2	1:2.5
	AS/FS	Semi-selective	Stacker rack	10080	12	$4\frac{1}{2}$	7	1.9	1:2.14
	AS/RS	Hi-density	P/FLO	12600	12	$4\frac{1}{2}$	3	1.5	1:1.71
	AS/RS	Hi-density	Push BK	12600	12	$4\frac{1}{2}$	3	1.5	1:1.71
80 ft	AS/RS	Selective	Stacker rack	11520	16	$4\frac{1}{2}$	12	1.7	1:2.5
	AS/RS	Semi-selective	Double deep	13440	16	$4\frac{1}{2}$	7	1.4	1:2.14
	AS/RS	Hi-density	P/flow	16800	16	$4\frac{1}{2}$	3	1.15	1:1.71
	AS/RS	Hi-density	Push BK	16800	16	$4\frac{1}{2}$	3	1.15	1:1.71

foot and cubic foot utilization between product and space. The chart shows that a high-density storage system, as opposed to a selective rack storage system, reduces the cubic storage to space ratio by approximately 50%, hence reducing high refrigeration costs.

Of course, with any given set of advantages there must be some disadvantages, and this is also true for pallet live storage. The only way to ensure constant operation is to ensure constant roller and pallet surface conditions. To do this, it is necessary to control the skatewheel track sections, pallet surface and control the pallet load velocity and direction:

- It is important to make sure that the pallet has a uniformly distributed load and that the centre of gravity falls within the pallet depth
- It is necessary to make sure that the pallet touches as many skatewheels as possible to distribute the load and keep the individual skatewheel loading to a minimum
- Free run of the pallet must be kept to a bare minimum in order to control the velocity and eliminate the possible overlapping of loads and spillage

Pallet   To assure high reliability of a pallet live storage system, it is recommended that a captive pallet be used with a specially formed metal runner. The metal runner is screwed to the pallets with self-tapping screws. Each pair of runners has a knurled running surface to improve braking friction and a raised rib to ensure tracking. This becomes extremely important when the pallet flow track is 100 ft (30 m) or more in length and several feet above the floor. The most popular pallet is a plywood board $\frac{3}{4} - 1\frac{1}{4}$ (19 mm–32 mm) thick. The average pallet with metal runners attached weights 45–50 lb (20 kg–23 kg).

Skatewheel track   With precision rollforming and a prepunching of a skatewheel track section, it is possible to guarantee that at least 80% of the skatewheels (1.9 in (48 mm) diameter) mounted on 2 in (50 mm) centres have pallet contact.

Therefore, a 48 in (1.2 m) deep pallet, 2000 lb (909 kg) capacity has:

$$\frac{2000 \times 100}{24 \times 2 \times 80} = 52 \, \text{lb} \ (23.6 \, \text{kg}) \ \text{per skatewheel}$$

The slope of the skatewheels is a function of pallet capacity and brake placement. Based upon the placement of brakes on 50 in (1.27 m) centres for a 48 in (1.2 m) deep pallet and a 2000 lb (909 kg) load, a flow lane will have a slope of 2% (1 in (25 mm) per pallet depth). These type of systems will accept a weight range of 2.5 to 1.

Braking   The speed retarder is the heart of a live pallet storage system. A rubber wheel mounted in a retarder unit replaces skatewheels at each pallet location. A friction wheel surface selected for operation at any temperature assures slip-free contact with the slave pallet metal runner. The retarder wheel is geared to an inertial control that ingeniously reduces excess flow speed without restricting even the start-up of a stationary pallet load. The retarder unit is virtually maintenance-free and wheel surface wear is minimal. A special compound is used for retarders that are used in freezers to eliminate seizures. The normal running speed of a pallet load is 30–50 ft per minute (9 to 15 m/min).

Entry guide Entry guides at each lane assist in centring the pallet load properly on the flow tracks with due concern for the equipment and the load. This also ensures improved productivity levels.

Pallet separator When retrieving unit loads with an automated stacker retrieval machine, it becomes necessary to separate the exit pallet from the rest of the line. The exit-end consists of a spring-loaded endstop connected to a pallet separator stop which is designed to create a 12 in (305 mm) gap between the first exit pallet and the rest of the loads in the lane. This minimizes pallet damage and holds the back pressure as the load lane fills. When the exit position load is removed the separator automatically releases the second load and captures the third as the line moves forward.

Push-back rack

The push-back rack uses the same components as the live pallet storage with only the crane shuttle mechanism being changed to accommodate the pushing forces required.

The mechanical efficiency of push-back rack systems is such that it takes only 90–120 lb (41–55 kg) of force to push a 3000 lb (1364 kg) pallet load up a 3% slope, therefore the lane depth is really only limited by the capabilities of the crane shuttle mechanism used. Stacker cranes are good for push-back as they are guided top and bottom and have sufficient mast stiffness to withstand the side forces. It is possible to push at least eight pallets deep either side from a single aisle. The use of a special slave pallet is recommended, the same type as used in deep lane flow rack systems.

Push-back rack systems are used where a high density of storage is required with few SKUs (i.e., ice cream, pizza) and the activity rate is low. A single crane aisle is all that is required.

Using a combination of push-back and flow rack eliminates the need to have static racks on the outside crane aisle to support the walls of a rack-supported building.

Rack-supported structures

All rack systems can be designed so that they become the skeleton for the installation of insulated roof and wall panels (Figure 11.6). The initial cost differential between a rack-supported system and a free-standing building with a free-standing system inside it probably does not vary greatly. The major advantage of the rack-supported system is that it can be installed and implemented much faster than the conventional building and free-standing rack. It is possible to pick up three months or more time on the project schedule as the system supplier does not have to wait for the building portion to be complete; he simply needs a cured concrete foundation poured to good close tolerances. Building trades then work in concert with the rack installer.

As the rack is designed to distribute the static and dynamic loadings over the whole concrete slab, the installation of evaporators and ductwork becomes less cumbersome. The rack structure normally has enough strength that the additional loading of the evaporators does not increase the cost of the rack structure. The evaporators can be installed in penthouses on top of the roof system, or inside the actual storage system with a series of internal catwalks installed to allow for preventive maintenance procedures.

It is recommended that all structures installed with automated stacker cranes have an expansion joint running across the entire width of the system. During the cool-down process bolted rack structures do have a tendency to contract, reducing running clearances and sometimes causing deformation to a structural support member. The expansion joint reduces the chances of this happening.

Figure 11.6 High density rack supported structure with insulated panels.

It is good practice to use chemically treated anchors instead of expansion bolts to fix the rack and crane rails to the floor. If an installer drills a hole and goes through the insulation the chemical anchor will seal the hole, hence keeping the freezer box airtight and reducing the possibility of the concrete slab heaving.

The rear runout end of the crane aisle should be made big enough to allow preventive maintenance personnel to move around the stacker crane and carry out their duties in the shortest possible time. Storage of crane spare parts within the freezer is also a good practice to establish. This will eliminate condensation build-up on a component stored in an ambient temperature storage location.

11.6 Delivery systems

Many different delivery systems have been used to feed a high-rise storage system and these systems can be as simple as a fixed pickup and delivery (P&D) stand or as complex as multilevel conveyor systems.

The fixed P&D station is the least desirable, due to the fact that it has to be fed by an FLT, requiring that a large door be opened and closed for every pallet movement into and out of the system. This causes air of different temperatures to meet, usually ending up with icing around the door seals and on the floor making the freezer unsafe and inefficient.

The most practical solution is to convey the product into and out of the freezer box using two small openings just big enough for the pallet load to pass through. In order to control and separate the air temperatures between two areas, the practice of using a vestibule with automatic sliding or guillotine doors is preferred and affords the most protection from ice build-up and snow. There are more economical solutions such as rubber strip curtains, or bi-fold doors activated by the movement of the pallet, but they are only a compromise. The added cost of the vestibule door will pay for itself quickly just in the refrigeration cost savings.

No matter what system is used to keep the freezer sealed, there are some basic rules which should be kept in mind when using conveyors on the front end of the system:

- Any pallet to be stored must be conveyable. Do not allow bad pallets to enter the system; they are difficult to get out once they are in
- All loads must be sized and/or weighed to make sure they fit the system design parameters
- A rework area for the rejected size loads must be accommodated outside the freezer
- If lift trucks are feeding fixed P&Ds then guarding should be used to protect the crane and rail
- Conveyors that are not at floor level must have access for maintenance as

a conveyor represents a single point failure area and can possibly shut the whole system down. Quick repairs are necessary

- Any interface with lift trucks must be very strong and be able to take abuse
- Make sure machine-readable labels are clear and in the correct position on the pallet or load
- Make sure conveyors can be manually operated in zoned areas

11.7 Automated storage machines

Stacker cranes, as they are universally known, were developed mainly for regular distribution warehouses along with in-process and raw material manufacturing facilities. The ideas and experience gained from these conventional situations soon found applications in the freezer, where many problems arose. Most of the problems were due to cheap designs and poor maintenance: As the frozen food warehousing industry increased in size, so the crane manufacturers started to learn more about the hostile environment of a freezer. Designs were changed to improve the ruggedness and reliability of cranes; and the situation has now turned full circle to the point where dry goods distribution centres are the recepients of these rugged reliable designs.

Basically two types of crane are manufactured today, single-masted and double-masted. The functions of the two are the same and vary only depending upon a particular manufacturer's standard. Some suppliers make cranes up to 100 ft (30 m) single mast, where others change over at 50 ft (15 m) to a double mast. 80% of all cranes manufactured are designed for the conventional grocery pallet dimensions of 48 in × 40 in (1.2 × 1.0 m) with weight ranges running from 1500 to 4000 lb (682 to 1818 kg).

All cranes can be manually driven with the addition of an operator's heated cabin, which is considered by many to be essential. The operator's cabin travels up and down with the load and carriage so that the operator has close visual control of all motions. A cab is recommended even under automatic mode for backup.

With the declining costs of automation, the majority of systems are now running under remote computer control. Cranes that do not have an operator's cab have the crane controls in a heated enclosure on the base of the vehicle. From this enclosure it is possible to run the crane in a semi-automatic mode or run all functions manually. The only disadvantage to this is that it is very difficult to store or retrieve loads at 60 to 80 ft (18 to 24 m) levels unless some of the automated location aids are used in conjunction with the manual controls.

Crane speeds are determined by the height and length of the aisles and the desired activity rates but generally cranes will run within the following ranges:

- Travel speed 300–500 ft per minute   (90–152 m/min)
  acceleration/deceleration 1.1 ft per second^2 (0.0335 m/sec^2)
- Hoist speed 60–120 ft per minute   (18–36 m/min)
  acceleration/deceleration 1.5 ft per second^2 (0.457 m/sec^2)
- Fork/extractor speed 60–120 ft per minute   (18–36 m/min)
  acceleration/deceleration 1.2 ft per second^2 (0.366 m/sec^2)

On-board controls

Automatic control is accomplished today using either an on-board PLC, or a small microcomputer communicating up to a remote warehouse control computer (WCC). Eventually all on-board control systems will be microcomputer-based. The on-board controls are divded into the following modules:

- Computer and input/output (I/O)
- Drives
- Sensors
- Safety logic

The on-board control unit which is housed in a panel and mounted on the rear of the crane contains the logic required to control all the movements of the vehicle.

The microprocessor is used to command the horizontal, vertical, and shuttle motors through each cycle. It receives commands from the WCC to store and retrieve loads. In turn, it generates the command signals to drive the motors to satisfy these requests. It continuously monitors its position data. It monitors for error conditions via the various safety sensors and should an error condition develop, it applies the necessary inhibitors. All error conditions that imply a potential hazard result in a power-down of all drive power.

The computer sends status information back to the WCC via the communication system so that it knows when the S/R machine is ready for another command.

Various control programs, error tables, and communication programs are resident in the EPROM memory of the on-board controller. These programs can be readily modified and new EPROM memory chips burned to provide altered capabilities. Program alterations can be accomplished on an IBM PC or compatible, with the new programs installed into standard purchasable EPROM memory chips.

Operational parameters are stored in non-volatile read/write memory and may be altered by trained technical personnel on-site using an IBM PC. These parameters may also be printed and stored on diskette files for backup or offline review.

The on-board microcomputer responds to commands in a fully automated mode. If errors occur during the execution of the commands, an error message is transmitted back to the host. Depending on the type of error, automatic recovery is possible. Hardware errors, load skew errors, time-out errors, etc., will require human intervention for correction of the error and a physical reset. On-board maintenance controls are provided for maintenance purposes and are not suitable for load storage/retrieval operation.

All signals brought into the computer are optically isolated for noise immunity and are appropriately filtered. An indicator LED is provided on the status panel to show the current state of all digital input and output signals. All digital input and output signals are wired to plug-in modules facilitating quick replacement.

The on-board microcomputer has a fan-cooled multibus back lane. The main CPU board is an 8088 CPU with three serial ports consisting of a computer interface, CRT/keyboard, and diagnostic display. The computer has a parallel port for sensor I/O and executes communication, I/O, and motion control firmware.

The encoder CPU board is an 8088 CPU with a parallel port for encoder and sync sensor inputs. The CPU executes encoder and sync sensor firmware.

The microcomputer has a battery-backed RAM board for non-volatile parameter storage and an analog output board that interfaces the microcomputer to the motor drives.

The I/O consists of one multiplexed I/O channel with the following inputs:

- Safety switches
- Drive fault circuits
- Load sensors
- Motion slow down and limit switches
- Conveyor interface eyes

and the following outputs:

- Motion commands
- Conveyor interface eyes
- One nonmultiplexed high-speed input channel for the travel and hoist encoders, along with travel and hoist sync sensors
- An I/O subsystem watchdog circuit which halts the system if the microcomputer is not functioning

Status of each I/O device is always available via LED indicators. Complete operational fault reporting and real-time validation of sensors is part of the diagnostics. A cyclic redundancy check is performed on all characters received from the WCC.

An initialization command to test end-of-travel sensors for safety is used. Position information and error messages are displayed on a diagnostic display mounted on the main electrical door.

As mentioned earlier in this section, cranes that operate in freezers may need special considerations during the design phase, which are as follows:

- All control enclosures must be heated and located close to the ground for ease of maintenance
- Motors fitted with low-temperature steel shafts
- Gearboxes equipped with double seals on all output shafts
- Bearings should be sealed and low-temperature grease applied
- Low-temperature oil in all gearboxes and hydraulic shock absorbers
- Low-temperature electrical wire and cabling
- Eccentric cam followers to maintain tight clearances on the crane rail
- All crane rails mitred at the ends to reduce wheel wear
- A lanyard cable down the length of each aisle to act as an emergency stop in case operating personnel are caught in an operating crane aisle is desirable
- Cranes when not in operation should not be parked next to a door where they could be prone to condensation and icing

All of these considerations are intended to reduce the amount of maintenance that has to be performed. It is good practice to make the preventive maintenance time periods shorter than those for ambient storage systems.

11.8 System controls

There are three major subsystems in the freezer automated warehouse. The automated storage and retrieval system (AS/RS) provides for storage and automatic retrieval of parts within the warehouse. The input/output (I/O) system provides the horizontal transportation of parts to and from the AS/RS and may include conveyors. The third major subsystem, the control system, links the AS/RS and the I/O system, scheduling and coordinating their operation while also managing warehouse information.

When control is effective, the benefits to the automated warehouse are numerous. For example, the requirements for labour, floor space, and energy are reduced. In the distribution (or finished goods) warehouse, specific benefits have included improved inventory accuracy, increased part number turns, and reduced inventory carrying costs.

To produce such benefits, however, the control system must be designed with certain requirements in mind. The major requirements that influence the design of an automated warehouse control system can be grouped as follows:

- Operations requirements (general and application-specific)
- Database and information flow requirements
- Requirements related to physical elements of the warehouse environment

In discussing these requirements, this section outlines the essential features of an automated warehouse control system. Drawing on actual system designs and user experiences, it discusses the requirements in both general and application-specific terms.

General operational requirements

Some operational features must be designed into the automated warehouse control system regardless of the warehouse's application. These common application requirements are discussed in the following sections.

System integrity safeguards   Most control systems for automated warehouses handle several thousand file transactions a day; in fact, a very active system will process four to six file transactions per second, or roughly 20 000 per hour. The control system designer must incorporate various safeguards to ensure that system integrity will be maintained during all of these transactions. Operational requirements related to system integrity fall into three categories:

- *Controlled start-up and shutdown.*   As they are with any on-line system, the start-up and shutdown sequences in the automated warehouse are crucial for maintaining overall system integrity. Start-up should not begin until initialization of the various pieces of automated equipment and of the on-line database files has occurred. Similarly, shutdown of the automated warehouse operation should not occur until the control system has ascertained the status of transactions and equipment operations.

   To accommodate this requirement, a special mode of control system operation called the *quiescent mode* is needed. During this mode, no new transactions are activated, and those transactions that are in process are allowed to run to their normal conclusion. The quiescent mode allows controlled stopping even in the event of critical component failure.
- *On-line operation status indication*   For an automated warehouse control system to be effective, on-line and real-time information must be available to the controls supervisor. Thus, the control system must report equipment failures and critical conditions so that appropriate action can be taken. This information could be conveyed by audible alarms, printed reports, or CRT displays.
- *Backup and recovery techniques*   To ensure database integrity during a critical path hardware failure, backup and recovery techniques must be incorporated into the control system. Three general techniques—cold, warm, and hot backup—are used.

   In a *cold backup* system, the application software runs an on-line audit trail by capturing a chronological record of the transactions on a disk or magnetic tape. Occasionally, the computer system is taken off-line and a copy of the primary files is made on a second tape or disk. This copy is called the *checkpoint file*. If the system fails, control of the automated warehouse devices must be physically switched from the primary computer to a backup computer. A computer program for recovery allows the audit trail information to be replayed or transacted against the checkpoint file up to the point of failure. That is, the computer acts as if the record that it is reading back is actually happening and rebuilds the active database accordingly.

A *warm backup* uses a duplicate or mirror image of the primary files. When a file is updated on the primary disk, the same update is processed onto a second disk. If one of these critical disk devices fails, the operational device can continue running in a warm, or continuous backup, state. Recovery simply involves copying the data file that was used during single disk operation to the new mirror file once it has been repaired and returned to service.

Hot backup is provided by two concurrently operating sets of computer hardware; as each transaction in the primary computer is processed, a duplicate transaction is processed in the backup computer. This requires complete redundancy of hardware and software. Some hardware vendors provide an automatic means for each processor to check the status of the other; then, when one fails, it is taken off-line, and the operation continues processing. In other cases, some operator intervention is required to switch external work stations and devices to the hot backup system.

System security safeguards   The physical security of the control system for the automated warehouse is attained through methods similar to those used for other computers; that is, through controlled access to secured computer rooms. Meeting the operational requirements for computer security in an automated warehouse involved restricting terminal access and providing operator tracing and accountability.

- *Restricted terminal access*   This security measure can be implemented in several ways. Some of the most commonly used methods are as follows:
  - Badge-reading devices that permit access when they read the employee number on a badge
  - A non-displayed access code that must be entered by the employee during the log-on transaction
  - Restrictions that limit the kind of transactions that can be performed on a specific workstation

  Sometimes all three security methods are used in a system. For example, the user gains initial access by inserting a badge into the badge-reading device or by entering a non-displayed access code (or valid employee number). Once on the system, the user can perform only transactions that are required at that workstation. A simple table can be used to define the allowable transactions at each workstation. The table can be easily modified as necessary
- *Operator tracing and accountability*   When tracing and accountability functions are required, the log-on transaction should require the entry of an employee number. This number would then be associated with every function processed at that workstation during its log-on period. The

audit trail estabilished by the backup system could be reviewed to provide accountability for questionable transactions.

System-generated reports   Six basic reports should be provided by the warehouse control system:

- *Location inventory report*   This gives status and other applicable part number information (e.g., description) in location sequence. The report includes all locations under the system's control domain, which often extends beyond the automated portion of the warehouse to conventional storage areas.
- *Material inventory report*   This is very similar to the location inventory report in information content, but it is presented in part number sequence.
- *Material requisition report*   To measure backlog, it is often necessary to print a report of all entered or received requisitions. This report lists requisitions in process as well as those awaiting processing in the system, but does not include those that have been processed. It may also be known as the *orders report* or the *pick list*.
- *System activity report*   This shows the activity of various key resources in the automated warehouse, such as the AS/RS machines, pick cells, conveyors, and AGVs. The report includes such information as the number of input transactions, output transactions, and machine errors; the amount of time on-line and off-line; the percentage of equipment utilization; the total number of transaction currently queued for processing; the number of transactions currently being served; and the available information on transaction queue sizes.
- *Automatic cycle inventory report*   This report lists the part numbers that have been requested to be cycle counted and those that have been counted. The report also compares count results against the database to reveal any discrepancies.
- *Error logging report*   This is a dynamic log that notes the date and time anomalous situations occurred. It is used to activate whatever processes or procedures are required to correct the anomaly.

External interfaces   Most automated warehouse control systems have an external interface to other control systems (e.g., the host computer that maintains financial data). This interface may be batch oriented, using cards, disks, or tapes, or—like most systems today—it may be direct telecommunications interface. A telecommunications interface does not necessarily mean that the two systems are on-line and communicate in real time. The telecommunications interface may be a batch process that occurs once a day or even weekly. The design of the external interfaces and their place in the overall operation are determined by the specific application.

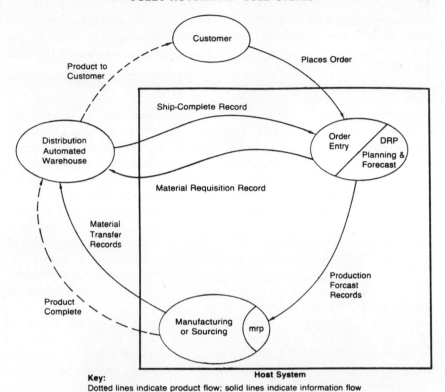

Figure 11.7 Common distribution warehouse operation.

Distribution warehouse requirement Figure 11.7 illustrates a common operation scheme for a frozen food distribution warehouse. This scheme is simplified to concentrate on the flow of customer orders and the key elements that are used in automated distribution warehouse systems.

The host computer generally performs the direct customer order entry functions because it has access to such financial systems as accounts receivable. The host system will also contain history and other important information, which feeds the distribution requirements planning (DRP) software. From the DRP software, a production forecast record is sent to the manufacturing systems, which in turn use material requirements planning (MRP) software to plan production of the product. As an order is received from the customer, a material requisition (MR) record is sent electronically to the automated distribution warehouse computer. The MR is processed against the on-line inventory and completed when the material is sent to the customer. A ship-complete record is returned to the host computer to trigger closeout of the customer order and to initiate the accounts receivable process.

Meanwhile, as manufacturing completes a new product, it is dispatched to the automated distribution warehouse along with a material transfer record.

This record is received at the warehouse before the completed product. On receipt of the product, the warehouse control system can easily identify the material using the material transfer record. However, this process is initiated manually.

This description of the operation scheme highlights two functional areas in which unique operational requirements for the distribution warehouse are most apparent: requisition processing and inventory control. Some of the unique requirements are shown in Table 11.4; to indicate relative importance, a 0–5 rating system is used where 0 indicates 'never used' and 5 indicates 'always used'. The intent of this chart is not to fully define each operational requirement, but rather to show how each requirement relates to a specific application segment. As such, the table can be used as a checklist to identify unique requirements during control system design.

Table 11.4 Operational requirements for the distribution warehouse. Numbers indicate relative importance on a scale of 1 to 5.

Unique requirements	Importance
Requisition Processing	
• Priority Activation	5
• Sequencing by:	
—Line Items to Order	1
—Order	4
—Carrier/Truck	4
—Stop	4
—Time	4
• Order Fill Algorithm:	
—First In, First Out (FIFO)	1
—Last In, First Out (LIFO)	0
—Best Fit	0
—Modified FIFO	4
• Ship Preparation:	
—Packaging	0
—Accessory	0
—Consolidation	4
—Ship Documents	4
Inventory Control	
• Shelf Life	4
• Activity Zoning	4
• Lot Tracking	1
• Service Parts Logical Segregation	0
• Quick Pick Location Replenishment	5
• Seasonal Surge	1

Key:
5   Always used
0   Never used

Database and information flow requirements

The relationships among database elements in the automated warehouse control system and the host system as well as the associated information flow between the systems depend heavily on the database design and the media on which it resides.

Five critical types of data must be maintained in the database of the automated warehouse control system. These areas are:

- Part number data
- Inventory data
- Transaction data
- Physical device (AS/RS, conveyor) status
- Requisition data

The host computer is usually designated the master for part number information (Figure 11.8). Hence, part number information and all changes relative to it will originate through transactions on the host computer. The master file contains many data elements that are not necessary for operation of the automated warehouse computer. A subset of the necessary master file elements can be sent to the automated warehouse computer to allow for syntax checking (e.g., number of digits entered, entry format) verifications, and inquiry and reporting functions.

The inventory master file is maintained by the automated warehouse computer because that computer is directly responsible for inventory integrity and for maintenance of inventory information by location. The transaction, physical device, and requisition files are also most appropriately maintained on the automated warehouse computer.

Additional files may be required for specific functions and features of the control system. Some necessary files might be:

- *In-transit inventory file* Used to trace product as it flows out of the automatic storage area.
- *Automatic cycle inventory file* Used to identify SKUs for cycle counting.

Requirements related to the physical elements

The physical layout of the facility and the automated material handling and storage equipment affect the design of the automated warehouse control system. The impact of several key physical elements on control systems design are discussed in this section.

Input/output systems The method of I/O to the storage and retrieval system is a key physical factor affecting the control system design. The I/O system transports materials to and from the AS/RS and incorporates automatic sortation capabilities where needed. Many types of horizontal transportation

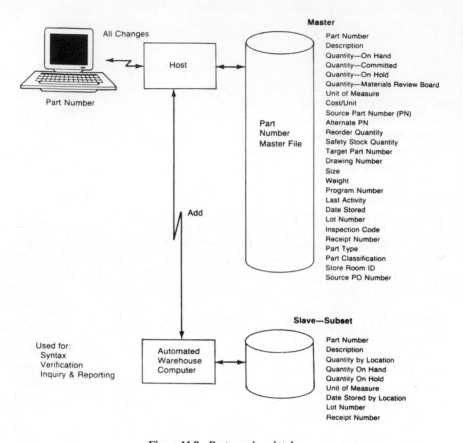

Figure 11.8   Part number database.

equipment are available for I/O systems, e.g.:

- Pallet conveyors
- Tote and transporter conveyors
- Sortation systems
- Tow-line systems
- Car-track/minitrain systems

Transportation systems have varying degrees of complexity and flexibility. Regardless of the system selected, a specific method of control is required in the automated warehouse. Each unit moving in the system must be assigned an operator—and machine-readable carrier identification. Logically associated with this carrier identification is information concerning source, destination, and part number. Thus, the operator–readable carrier identification can also serve as the external key for correcting anomaly situations.

Another key to a successful automated warehouse control system is separation of transportation system control from the management-level warehouse computer. This is done through the creation of a simple interface between the transportation controller and the warehouse computer. The warehouse computer sends the transportation controller a 'move' command, which includes information about the product to be moved and the locations the product is to be moved to and from. The actual moving and controlling of that product from the source to destination becomes the responsibility of the transportation controller. Upon successful completion of the 'move' command, the transportation controller responds to the automated warehouse computer with a status code indicating the disposition of the move request.

Computer and communications hardware The type and configuration of computer and communications hardware greatly affects the performance of this automated warehouse. The automated warehouse typically performed at each level. For large systems, an additional level between three and four, called a cell controller, is common. This additional level is necessary to coordinate timing-sensitive operations, such as robotic work positioning and precision I/O conveyors. This additional level also serves as a multiplexor of several local device levels, feeding back to an area controller such information as service status. Another advantage of this cell level controller in large systems is that it provides greater functional flexibility when changes are required in a related set of functions and operations.

Environmental conditions Environmental conditions that affect the control systems design are usually related to the computer hardware and associated peripherals. In general, the warehouse computer and its corresponding peripheral devices and backup systems should be located in an environmentally enclosed computer room that has appropriate access and security safeguards. Remote controllers (e.g., for I/O systems) can be positioned on the warehouse floor for ease of debugging and maintenance. Controllers that require data or disk operating systems should be placed either in an environmentally controlled enclosure or in a computer room or business office.

11.9 System justification

Each frozen food warehouse or distribution centre has its own set of unique requirements and location idiosyncrasies making justification a challenge. As mentioned earlier there are many intangibles connected with automation where arbitrary values must be imposed and accepted through each level of management approval. The bottom line is that the system be cost-justifiable in preference to other conventional methods, with a return on investment of 3–5 years or better.

Buildings

The storage system space allocation chart shows the relative area differences for different types of systems in numbers of pallets stored within a 19 200 square foot (1784 m²) warehouse. In a 20 ft (6.1 m) high building using 100% selective pallet rack four storage levels high using counterbalanced FLTs it is possible to store 1920 pallets. A 60 ft (18.3 m) high building using 100% selective stacker rack storing 12 levels high with narrow aisle stacker cranes can store 8640 pallet loads. This is an increase in storage capacity of 450%.

Generally AS/RS systems 60 ft (18.3 m) high will use 70% less floor space than a low-rise FLT system. Hence building costs for site preparation, foundation, utility equipment, sewers, and land costs are greatly reduced. As mentioned earlier in this chapter, it is necessary to determine the type and height of the storage system in order to calculate these cost savings. If expansion is an immediate requirement there may be an extra calculation to do to compare the savings of leasing off-site storage facilities.

Material handling equipment

The cost of stacker cranes and racks will be much higher than the conventional FLT and rack system given the same storage capacity and activity rates, but this difference has to be compared to the reductions in manpower and operating costs. As a rule of thumb one automated stacker crane does the work of two FLTs but has a much longer life expectancy.

Auxiliary equipment such as conveyors, pallet collectors/dispensers, weighing and sizing stations (Figure 11.9), must also be compared for labour and operational cost savings.

AS/RS systems are generally designed to last 15 years with good preventive maintenance programmes whereas an FLT system generally lasts 5–7 years before the operating costs become a liability and equipment is replaced. If may be necessary to replace FLTs 2–3 times during the life of a stacker crane, and each time an inflation cost must be included.

Modular AS/RS systems can be expanded very easily and controls can be upgraded to current specifications if further savings can be realized by adding other functions.

Operating costs

Direct/Indirect Labour Direct labour savings of 50% or more can be achieved from automating the following functions:

- Receiving
- Identification/inspection
- Transportation/dispatching
- Storage/retrieval

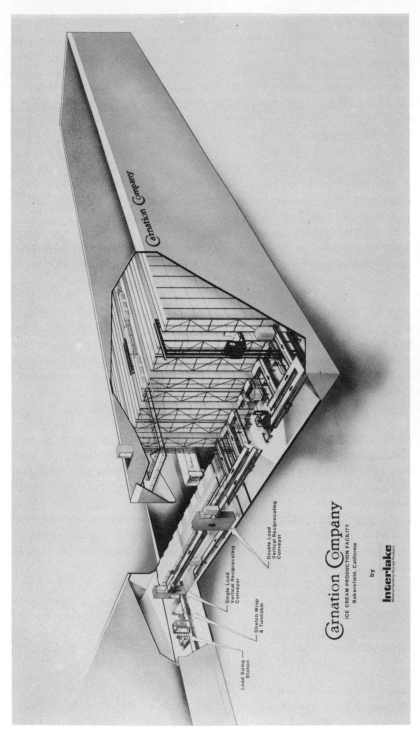

Figure 11.9 Example of an AS/RS with auxiliary equipment.

- Order picking
- Sizing/weighing
- Shipping

These savings are magnified when the temperature is $-20°F$ ($-29°C$).

Savings in these areas also have an impact on office/clerical help, especially if real-time computer control is being used. Support personnel for inventory and order scheduling can be greatly reduced.

Inventory carrying cost   Through improved record keeping and scheduling it is possible to reduce inventories by 10–15%. The problem of misplaced inventory due to bad record keeping can be eliminated by using a discrete storage slot location address system.

The annualized cost of carrying inventory can be as high as 40% of the inventory value. The most efficient and profitable freezer warehouses are those that turn 26 or more times per year. This is attainable in the grocery and frozen food industries, but not in the metal-working electronics industries.

Maintenance   Equipment is designed today to run at efficiencies of 98% or better. This is due to standardization of product, plug-in components, and built in on-board diagnostics which pinpoint the problems, hence reducing the time to get the equipment back on-line.

Crane systems usually have a backup semi-automatic and manual mode of operation, so that under adverse conditions the equipment can store or retrieve product without relying on the computer interface.

Product damage   Industry figures show that product damage can run as high as 2%. Human error in guiding FLTs accounts for thousand of dollars in both product and equipment damage yearly. Mechanically guided cranes do not hit loaded pallets during store/retrieval functions and therefore reduce damage to almost zero.

Pilferage   The features that reduce product damage serve equally well to reduce pilferage substantially. Product that is stored 60 ft in the air with no access other than by an authorized warehouse operator cuts down any incentive to pilfer. If there is a loss it is much easier to pinpoint the culprits. Voice recognition and thumbprint signatures are now available to secure a system from outside interference.

Cost justification

AS/R systems always cost more than conventional manual systems, and so it is always necessary to cost-justify the differential expense of the AS/RS.

Tables 11.5 and 11.6 compare a conventional 30 ft (9.1 m) drive-in warehouse with an automated, rack-supported, gravity flow system operated with

Table 11.5 ROI calculation AS/RS system ($000).

	Conventional	AS/RS	Better/ (Worse)
Capital cost differential			
Land:			
square feet	50 000	25 000	25 000
cost per square foot	$1	$1	
total land cost	50	25	25
Building:			
square feet	40 000	15 000	
cost per square foot	30	55	
building cost	1 200	825	375
Site preparation	320	210	110
total building cost	1 520	1 305	485
Equipment:			
conventional rack	285		285
AS/RS		3 100	(3 100)
fork trucks	350		(350)
fire protection	30	60	(30)
total equipment	665	3 160	(2 495)
Total investment	2 235	4 220	(1 985)

Table 11.6 Savings in annual operating costs.

	Conventional	AS/RS	Better/ (Worse)
Labour			
total headcount	14	4	10
annual labour cost	25 000	25 000	
Total labour costs	350	100	250
Maintenance	50	20	30
Utilities	100	40	60
Product damage	414	50	364
Pilferage	0	0	0
Insurance	0	0	0
Other	0	0	0
Total operation costs	914	210	704

two cranes 65 ft (19.8 m) high. Both systems store 3700 loads, 95 in high (2.4 m) 3300 lb (1497 kg) capacity.

Tables 11.7 and 11.8 show the cash flow statement and tax depreciation over a 10-year period showing that a 25% return on investment is possible, proving in this instance that an AS/RS frozen food warehouse is a very viable solution.

Additional operating costs that are intangible factors and could help swing the payback period to even better figures are:

- FIFO stock rotation
- Lost sales due to errors
- Periodic full physical inventory cost
- Poor customer service resulting from inaccurate stocktaking

In conclusion, AS/RS systems are relatively new but have proven themselves in hundreds of installations throughout the world in a very short period of time. The interface of equipment with low-cost computer control and bar code scanning, tied in with accurate inventory control has made it all possible. As technology improves, more and more pieces of the manufacturing and warehousing puzzle will be combined together to expand the complexity of freezer automated storage systems.

However, the final measure of acceptability for an AS/RS is the financial evaluation. Management must understand the advantages and then judge the factors of capital outlay, payback, the rate of return on investment against the gains, productivity, efficiency, and profit.

11.10   The future

With the popularity of microwaves and frozen food, the need for more freezer warehouse space will continue to increase rapidly. The choice of products available to the consumer will also continue to increase. This means that there are many challenges for the freezer warehouse management and the material handling suppliers.

The introduction of bar code labelling on all products has eliminated the need for human identification of products due to its high level of reliability. Bar coding will become more extensively used to track product from the manufacturing floor, through the distribution network, to the supermarket shelves.

As the number of SKUs increase, manufacturers will be manufacturing products in smaller lots, requiring a need for greater selectivity and the shipping of less than pallet loads to the distribution centres and stores. This will require that more automated picking be undertaken within the freezer. Order pickers will be replaced at some time by inexpensive robots, travelling along fixed picking rack faces, selecting cartons and placing them on to a takeaway conveyor sorting system which will move the product directly in to the shipping trucks. The freezer is a harsh environment in which to operate

Table 11.7 Schedule of tax depreciation.

	Cost	Year 1	Year 2	Year 3	Year 4	Year 5	Year 6	Year 7	Year 8	Year 9
Conventional										
land	50	0	0	0	0	0	0	0	0	0
building	1 520	46	48	48	48	48	48	48	48	48
equipment	665	95	163	116	83	59	59	59	30	0
Total depreciation	2 235	141	211	164	131	108	108	108	78	78
AS/RS										
land	25	0	0	0	0	0	0	0	0	0
building	1 035	31	33	33	33	33	33	33	33	33
equipment	3 160	451	774	553	395	282	282	282	141	0
Total depreciation	4 220	483	807	585	427	315	315	315	174	33
Incremental depreciation		342	596	421	296	207	207	207	96	(15)
Depreciation rates										
building	31.5 yr.	3.04%	3.17%	3.17%	3.17%	3.17%	3.17%	3.17%	3.17%	3.17%
equipment	7 yr.	14.28%	24.49%	17.49%	12.49%	8.93%	8.93%	8.93%	4.46%	0.00%

Table 11.8 AS/RS Justification: statement of cash flow ($000).

	Year 1	Year 2	Year 3	Year 4	Year 5	Year 6	Year 7	Year 8	Year 9	Year 10
Initial investment	(1 985)									
Operating cost savings	0	704	704	704	704	704	704	704	704	704
Higher tax depreciation	0	(342)	(596)	(421)	(296)	(207)	(207)	(207)	(96)	15
Pretax earnings	0	362	108	283	408	497	497	497	608	719
Taxes @ 34%	0	(123)	(37)	(96)	(139)	(169)	(169)	(169)	(207)	(244)
Net earnings	0	239	71	187	269	328	328	328	401	475
Add back non-cash expenses tax depreciation	0	342	596	421	296	207	207	207	96	(15)
Net cash flow	(1 985)	581	667	608	565	535	535	535	497	460
Cumulative cash flow	(1 985)	(1 404)	(737)	(129)	436	971	1 506	2 041	2 538	2 998

within productively, but at the present time there is no economical alternative for food storage.

Automated stacker cranes equipped with turret FLTs will enable the use of low-cost pallet rack at 80–100 elevations. This will allow warehouses to use standard GMA (Grocery Manufacturers Association) double-faced pallets and conventional pallet racks, eliminating the need for slave pallets and double handling. The rack cost will be reduced significantly, thereby lowering the initial investment.

New techniques of controlling stacker cranes at speeds 100% higher than those of today are being developed to give higher throughput rates per crane and allow looser toleranced rack installation. This means faster, lower-cost installation.

At the present time cranes have a 10–15 year life span. Control systems become obsolete quickly and expensive to retrofit to newer methods. Cranes manufactured in the future will have modular plug-in control system enhancements which will be easily adaptable to fit the customer needs.

Automated guided vehicles (AGVs) are becoming increasingly popular but have not yet reached the frozen food warehouse. This is mainly due to the shortened life expectancy of the batteries at freezer temperatures. Physically there is no reason why they should not replace the conveyors on the front end of a system. As battery technology improves and 'opportunity' charging becomes commonplace, the AGV will make a significant move to replace conveyors. This will improve the reliability of the total system. AGVs that need maintenance can be taken out of service and replaced, without disrupting product flow. This tends to give a system more flexibility. With only an embedded floor wire, the AGV can be rerouted to different locations, and vehicles added as dictated by demand.

Index

Cold and Chilled Storage Technology